THE
GENIUS
LIFE

ALSO BY MAX LUGAVERE

Genius Foods: Become Smarter, Happier,
and More Productive While Protecting Your Brain for Life

THE
GENIUS
LIFE

Heal Your Mind,
Strengthen Your Body,
and Become Extraordinary

MAX LUGAVERE

HARPER WAVE

An Imprint of HarperCollins*Publishers*

THE GENIUS LIFE. Copyright © 2020 by Max Lugavere. All rights reserved. Printed in the United States of America. No part of this book may be used or reproduced in any manner whatsoever without written permission except in the case of brief quotations embodied in critical articles and reviews. For information, address HarperCollins Publishers, 195 Broadway, New York, NY 10007.

HarperCollins books may be purchased for educational, business, or sales promotional use. For information, please email the Special Markets Department at SPsales @harpercollins.com.

FIRST EDITION

Library of Congress Cataloging-in-Publication Data has been applied for.

ISBN 978-0-06-289281-2

20 21 22 23 24 LSC 10 9 8 7 6 5 4 3 2 1

Dedicated to my mom, Kathy. I love and miss you.

CONTENTS

PREFACE

On the surface, my mother seemed to have checked all the boxes for a long life and great health. She wasn't overweight, didn't drink, and never smoked. She consumed lots of fruits and vegetables and always loaded up on low-fat and salt-free "heart healthy" grain products. So it was a shock to my family when, in 2010, at the age of fifty-eight, her brain began to fail her.

It was subtle at first but became apparent when we'd cook together—one of our favorite things to do—that her ability to perform simple tasks had become mentally labored. I'd ask her to pass a ladle, for example, and it would take her a few extra seconds to respond. It was strange to see my mom suddenly struggle, but no one in my family had ever had brain problems. I thought I was just watching her get older.

Things got a little more serious when she told the family that she had seen a doctor in New York, but even then, the details of her visit were murky, lost in translation between the fear and confusion she was probably experiencing. In August of 2011, we decided to book a trip to the Cleveland Clinic in Ohio and that I would accompany her. After running a battery of esoteric tests, the neurologist looked up from his notes and assigned my mom a strange Parkinson's-like disease. Handing us a few prescriptions, he sent us on our way.

Later that night, I did what any millennial with a Wi-Fi connection would do and consulted the oracle of our time: Google. I learned that my mom had been prescribed drugs not only for Parkinson's disease, but for Alzheimer's disease as well. *Why Alzheimer's disease?* I wondered. Did this mean my mom was going to die? Or forget who I was?

As these questions began to circle my head, feelings of fear and helplessness started to bubble up and spill over like water boiling on too high a flame. My heart started pounding, the room went black, and all I could hear was a ringing in my ears. I was having a panic attack. How could this have happened to the person I loved most, and right under my nose? What could be done? How could I save her?

The next day, we flew back to New York and began making more doctor's appointments. I went with her to all of them because if there's anything a career in journalism has taught me, it is how to ask questions. Desperate for answers, what we'd get was often little more than "diagnose and adios." Many times a doctor would add on a new drug to my mother's regimen or increase the dose of one she was already on.

Despondent but still hopeful, we kept searching. I did more research, we scheduled new appointments. And my mom always had a great attitude. "I'm happy that I've gotten so far," she'd say.

In the years that followed, my mother's symptoms grew worse, especially where thinking was concerned. Alzheimer's disease makes a person's memories ephemeral, like chalk on a sidewalk. In my mom's case, it was more like a slow and crippling strangulation of her brain power. She lost the ability to communicate with any depth or richness, and many times she'd lose her train of thought soon after starting to speak.

Her eyesight suffered. I'd catch her reaching for objects that weren't there, or "missing" what she intended to pick up. Reading was one of her favorite hobbies (my mom loved to collect books), but she was no longer able to do it. She struggled with basic everyday habits of self-care, "forgetting" how to use the toilet, feed and wash herself, and pick up the phone. Even opening doors became a challenge. And, of course, she could no longer leave the house alone.

Then there were the movement problems. My mom gradually became less and less mobile, suffering from weakness, stiffness, and instability. She came to rely on me, her health aides, or my brothers to sit down, stand up, and everything in between.

I was tasked with filling her pillbox, which at one point had nearly a dozen different pharmaceuticals in it. Though they were meant to help, the medications never seemed to do anything but make her feel worse. I'd often catch myself staring at the pastel-colored tablets, wondering how they each were interacting in her increasingly frail system. Giving them to her, I'd often feel like I was deceiving her. But what choice did I have?

It was Labor Day 2018 when everything changed again. I was in Los Angeles for work when I received a call from my brother.

"Mom is in the emergency room," he told me.

"For what?" I asked. I was with her only a few days before and had taken her to the doctor. There had been a decline in both her appetite and cognition, but the doctor's visit was, as usual, frustratingly unremarkable.

"She turned yellow," my brother told me. Confused and worried, my family had rushed her to get help.

"Well, what's wrong?" I asked.

"They don't know," he said. "They think it could be a gallstone, but . . ."

Before he could finish his thought, I hung up the phone and changed my flight to get the next plane out. "What now?" I wondered, anxious the whole way home.

When I arrived at the emergency room the next day, my mom was unintelligible and, indeed, a bit yellow. The doctors had just done an MRI of her abdomen.

A gallstone would have been a perfectly good explanation for her unusual hue, but what they found was much worse: a tumor. It was on the head of my mom's pancreas, pressing into her bile duct. This had caused bilirubin (the pigment that gives stool its color) to back up into her blood, seeping into her skin and eyes. And it looked like the cancer had already spread.

They put a stent into her bile duct, and we went home. It took a day or two for her color to return to normal. And her cognition improved right away. For the next twelve hours, she seemed almost like her old self. That night, my whole family present, we ordered Chinese food and put her favorite band, the Rolling Stones, on the TV.

The following three months, however, were wrought with pain, weight loss, and desperate attempts to find a cancer treatment that would buy time for my mom. There was infighting in my family as to how aggressively we should seek medical treatment. After trips to three different hospitals, it became clear that doctors were unable to offer her much, reminding me of all those early trips to neurologists' offices. And all my mom wanted to do, it seemed, was stay at home.

At 11 a.m. on December 6, 2018, at the age of sixty-six, she passed away. I, my two brothers, Andrew and Benny, and our dad beside her.

My Hope for You

My mother's health suffered immensely, and watching her lose everything was heartbreaking. Was there anything that could have been done to avert her illness? What factors were to blame for her transformation from a seemingly healthy person to someone ravaged by disease? What could I do to improve *my* prospects for a long life with sound body and mind? Questions like these have become my obsession.

Although I began in the desperation of a family crisis, my search for answers has taught me more than I ever could have imagined about human health, and especially the brain. I've had the privilege to learn from scientists at top research institutions around the world, and I've been able to collaborate with many as well. I've created educational tools used to teach the clinical practice of dementia prevention, and I've even coauthored a chapter in a medical textbook on the topic.[1] And my discoveries about the important connection between our brains and what we eat were the basis of my first book, *Genius Foods*. Since its publication in 2018, I've received thousands of notes from doctors, nurses, dietitians, and nutritionists all around the world, many of whom have recommended it to their patients.

The work I did on *Genius Foods* changed my perspective on diet, but nutrition is a continually evolving science—and it's just one part of the puzzle for optimal health. So, in mid-2018, I launched a podcast, also called *The Genius Life*. Through it I've had the ability to learn even more about the brain-body connection from researchers working on the frontlines of nutrition and fasting, circadian biology (which examines our body's relationship with time), sleep science, exercise physiology, and more.

For a long time, people believed that their genes were their destiny. And our genes do matter, but how much we can attribute our health problems to them is up for debate. In the United States, the most well-defined Alzheimer's risk gene, carried by one in four people, increases the risk for Alzheimer's disease anywhere between two- and fourteenfold. In other parts of the world, that same gene has little impact.[2] Many cancers are thought to be triggered by our environment as well and are becoming more common. A woman's lifetime risk of developing breast cancer, for instance, was about one in twenty in the 1960s. Today, it is one in eight.[3]

Over the last seventy years, our genes haven't changed, but our environment has—a lot. New research is emerging all the time that points to the key role that environmental factors play in your health. Everything from the temperature and lighting of your home to the utensils you use to cook your food to the chemicals used to make your furniture are having a profound effect on your health and how you feel—and you likely *aren't even aware of it*.

Here's the truth: living in the modern world isn't good for us. Our bodies have defenses, but they can only contend with so much.

For starters, the food we eat is making and keeping us fat and sick. Being overweight is a driver of about 40 percent of modern cancers, and having an oversized waistline also correlates to accelerated brain aging.[4] All told, one in five deaths globally today occur due to diet alone, according to research published in the *Lancet*.[5]

But the food supply isn't the only problem in our fight for healthier bodies and brains. Our nights are now flooded with light, causing confusion to our internal timekeeping systems. We are deprived of clean air, sunlight (and the vitamin D our bodies badly need), and the many benefits associated with being in nature. Time spent exercising has declined dramatically while time spent commuting in cars or trains or sitting in front of TV screens has climbed sharply. Our homes have become saturated with untested industrial chemicals that wreak havoc inside us. We suffer from an epidemic of stress and few of us get enough sleep.

These forces gang up, overwhelming our bodies and leaving us anxious, depressed, and unwell. What's worse, we've come to believe that it's normal to be tired all the time; that chronic stress, anxiety, depression, and distraction are par for the course; that feeling bloated, fat, and weak is somehow how we're meant to feel. This state of affairs isn't natural, however. It's shortening our life spans and robbing us of our loved ones. And when it becomes unbearable enough, we self-medicate with food, drugs, reckless behavior, and, worst of all, apathy. But it doesn't have to be this way.

The good news is that many of the environmental factors that are making us sick fall under our control. We can restore our health by reengineering our habits and our habitats to resemble the environment in which human beings have evolved to thrive. This is what I call living a Genius Life, and it's accessible to everyone.

In the words of John F. Kennedy, the time to repair the roof is when the sun is shining. I was shocked to learn that dementia often begins in the brain *decades* before the first symptom. Even Parkinson's disease reveals itself late: by the time you experience your first symptom, at least half of the associated brain cells are already dead.[6] In truth, *none* of the conditions we fear most—including cancer or heart disease—develop overnight. To stand any chance at beating them, you *must* be proactive about your health. I wrote this book to help you build a strong, healthy, and resilient body, laying the groundwork for better health today, and for years to come.

THE
GENIUS
LIFE

INTRODUCTION

Have you ever experienced anxiety, fatigue, or brain fog? What about sluggishness, memory problems, or despair? Today these feelings are commonplace, but they needn't be. After all, we live in a time of unprecedented insight into the workings of the human brain, with millions of dollars spent annually to uncover truths about our most powerful organ. But these insights often go unnoticed, leaving many of us to suffer in silent desperation.

This book is going to change that. In the following seven chapters, you're going to discover powerful strategies that can help your brain function the way it ought to—not just the way you've accepted it to. That means less sluggishness, anxiety, and depression, and more energy and a better memory. If followed routinely, these strategies will even reduce your risk for some of humanity's most feared conditions—Alzheimer's, cancer, and heart disease, to name a few.

The gateway to healing your brain is your body. Myriad variables can affect the latter, and as a result can influence your thoughts, behaviors, and even emotions. New research, for example, shows the same steps that will produce a strong and resilient heart also benefit the brain. And that by losing weight or gaining muscle, we fortify our ability to feel happy and remember things. These relationships will become increasingly clear in the pages to come.

Another relationship to nurture—and one that has been most sorely neglected—is that of your body to its environment. We've inherited a world that is much different than that of our ancestors, which was the world in which our bodies and brains were designed to thrive. Twenty-first-century life and its trappings—including easily attained convenience foods, endless streams of digital distraction, and certain chemicals to which the average person is routinely exposed—all impose a burden on the body, overwhelming its defenses and predisposing us to sickness, malaise, and even shorter lifespans.

Mending these relationships is key to living a Genius Life. The following pages will endow you with a deep understanding of what it takes to get healthy, revealing the small changes that you can make in your daily routine that will have a huge impact on how you feel today, and your health later on. No matter your age or background, the time to act is now, and the plan is in your hands.

The Relationship Has Hit the Fan

For almost a century, experts believed that the brain was shut off from the rest of the body. Though connected to a blood supply, the brain once appeared to reside in a fortress of strict solitude, guarded by a checkpoint of cells known as the blood-brain barrier. But research over the past few decades has dismantled that idea. A new picture is emerging, one that proves that the brain and body are linked in innumerable ways.

Unfortunately, modern living does a number on the body. Many people are now overweight, and one in two adults have type 2 diabetes or are on their way there. Either of these scenarios affect an invisible process called metabolism, which is how our cells create energy. Problems with metabolism are overwhelmingly common. In fact, researchers are hard-pressed to find people in ideal metabolic health today.[1] And when metabolism falters, our brains begin to suffer.

If our cells can't create enough energy, inflammation ensues, which is a sign of immune activation. This also occurs when we're sick, and it's no wonder so many of us feel crappy all the time. Animals, for example, display striking behavior changes when inflamed. They lose interest in grooming and socializing, and eating, one of the most powerful instincts a creature has, becomes an afterthought. Human beings are no different. New research shows that anxiety, anhedonia (a decreased ability to feel pleasure), and mental fatigue can all follow blood markers of inflammation.[2]

Even depression may be a response to inflammation. It may surprise you to hear that one third of clinically depressed people respond poorly to traditional treatments but react well to anti-inflammatory drugs.[3] These drugs, often prescribed to treat bodily pain, seem to treat emotional pain for a subset of patients—a revolutionary finding in the field of psychiatry. Though the science is evolving, one thing seems clear: reducing inflammation in our bodies leads to happier, healthier brains.

One powerful lever at our disposal is food. Today our diets have become saturated with packaged, ultraprocessed products. These products now comprise 60 percent of our energy intake every day. Densely packed with empty calories and inflammatory chemicals, these foods are hard to avoid—they're so convenient and tasty!— but we must end the addiction if we want our bodies and brains to flourish. In chapter 1, I'll lay out a plan for eating the right foods for nutrient density and fullness, helping you shift your body into a healthier state without forsaking the pleasure of eating.

Having an optimally functioning brain and body requires more than just proper nutrition. For one thing, research suggests that when you eat may matter as much as what you eat. An ancient timekeeping system, hard-coded into your genes, regulates your body's key defense forces. Modern life has caused this clockwork to get out of sync, and this may be a factor in conditions like heart disease, cancer, and even dementia. Chapter 2 explores the rapidly evolving field of circadian biology, which has much to teach us about how to set our clocks so that we have optimal energy, focus, clarity, and digestion.

Our disconnection from the natural world is another source of trouble. Your doctor might not say as much, but the sun has medicinal value, influencing every organ in your body by way of the vitamin D it helps us produce. Unfortunately, most of us don't get enough sun. In chapter 3, I'll share how to get (and optimize) your vitamin D, ways to minimize the harm from air pollution, and how your ambient temperature can modulate fat burning, mental health, and even your risk for Alzheimer's disease and other forms of dementia.

Our thoughts may dictate our actions, but movement can change our thoughts. Getting fit is simply one of the best gifts we can give to our brain, strengthening its ability to stay resilient against degeneration as we age. For our premodern ancestors, exercise was a part of daily life. But modern life has all but eliminated our need for exertion, further advancing the metabolic mayhem. In chapter 4, you'll discover how to build strength and muscle, speed up your metabolism, and clear out all but the most stubborn of mental cobwebs with exercise. I'll include specific guidance for beginners and advanced trainees alike.

Unfortunately, some aspects of modern life are simply unavoidable; exposure to damaging industrial chemicals is one of them. History is littered with toxins foisted into the marketplace before appropriate testing has been done. Some examples include environment-destroying pesticides, lead-based paints, asbestos home insulation, and hormone-scrambling antiseptics in our soaps and toothpastes. A tiny fraction of these poisons may end up on the nightly news, but it's usually only after profound consequences to our health and environment have been wrought.

Between heavy metal overexposure and myriad toxins now present in everything from food containers to our furniture and even dental floss, your body is under attack. Some of these chemicals you've encountered every day of your life; the Environmental Working Group has identified over 287 industrial pollutants in the wombs of mothers-to-be, including pesticides, flame retardants, and waste from burning coal, gasoline, and garbage, each having clear links to neurotoxicity, developmental problems, and even cancer.

Prepare to be alarmed by what you read in chapter 5, but also to be empowered to identify and drastically cut down on your exposure to potentially hazardous chemicals. I'll dive into your medicine cabinet to reveal the drugs that may be doing more harm than good, and then into your kitchen to examine the dangers of common food additives and storage methods. Plus, you'll learn about heavy metals such as lead, cadmium, and arsenic, and what you can do to reduce your exposure and help detoxify what you've accumulated already.

Lastly, stress. Nowhere is our collective brain drain reflected more powerfully than in our stress levels, which soar to new heights every year according to the American Psychological Association. In chapter 6, we'll look at different ways of de-stressing, from finding meaningful work to managing your social media addiction (hey, we all have it), plus a meditation technique and mind-set tactics to help you better relate with those around you—including yourself. And, I'll share proven strategies to help you get the best, most rejuvenating sleep of your life, regularly.

I've written this book to be read from start to finish, but feel free to skip around. And if the information presented feels overwhelming, fear not; in chapter 7, I present the Plan, where I reintroduce many of the core concepts and overlay them onto your day.

Refresh, Reboot, Rebuild

Beginning with the first appearance of life on Earth approximately four billion years ago, every species has had to contend with various selection pressures, allowing only the fittest to pass on their genetic material. Over time, each generation would become strengthened by their natural habitat. All organisms—from the magnificent blue whale to the microbes in your mouth—are elegantly adapted to survive the environment in which they have evolved. You are no different.

Unfortunately, the last few hundred years have seen our environment mutate rapidly, and it continues to do so at a blistering pace.

Not all of the change is bad. We don't have to live in darkness when the sun goes down, the industrialization of our food supply means starvation is less of a threat than it was in the past, and who can argue with toilet paper? But the changes are too numerous and are occurring too fast for our bodies to effectively adapt and respond. As a result, we are fat, stressed, tired, and sick. But we need not be.

What we now know about the brain is that it is influenced by our bodies, which, in turn, are affected by aspects of our environment over which we have some control. That means that by following the protocol outlined in this book, you'll have a fighting chance at preventing the onset of major medical woes such as cancer, dementia, and autoimmunity. And unlike traditional medicines that target only single chemicals or biological pathways, the approach I've outlined considers the entirety of the system, strengthening your body in all of its miraculous complexity.

My goal in writing this book was to make the science of living long and healthy, easy and actionable. Whether it's your relationship with food, nature, light, stress, sleep, or even those nasty industrial chemicals I mentioned earlier, you now have a tactical and easy-to-follow guide to living your best life. I can't change what happened to my mom. But thanks to her, the rest of us have the opportunity to optimize our physical and mental well-being for a longer, healthier existence. The Genius Life is within reach, and I'm excited for you to begin this journey with me.

1

DON'T FORK AROUND

"Truth is singular; its 'versions' are mistruths."

—FROM DAVID MITCHELL'S *CLOUD ATLAS*

You might not get to choose your genes, but what you put past your lips is a decision that falls largely under your control. For that reason, food is a frontline defense against the conditions of aging and the foundation to living a Genius Life. The foods you eat can either feed or fight lethargy, malaise, and your predilection for disease. Food also dictates the amount of fat that you carry around with you on your thighs and waist.

There tends to be a lot of confusion about what we should—and shouldn't—be eating. Much of it is warranted. Studying diet is harder than studying drugs, and yet nutrition is much less well funded. Even among the scientifically literate, finding answers can feel like shooting a moving target . . . in the dark. Our healthcare providers are undertrained on these topics, and yet health books, documentaries, and celebrities alike all publicly proselytize their versions of the truth as if it's gospel. Then there's the media, which often resorts to sensationalized hyperbole—reliable as a traffic generator in the era of social media, but less helpful if you're actually looking for guidance. It's no wonder the average person is left feeling lost.

And yet truth and transparency are more important than ever. Food is as plentiful as it's ever been, but the majority of us are cripplingly undernourished. Corporations regularly launch food products to market aimed directly at the ignorance and fragile willpower of the average consumer. And industrial toxins are ever-present, from the oils used to cook our food in restaurants to the very receipts we are handed at checkout (innocuous as they seem, cash register receipts are usually covered in hormone-disrupting chemicals that feed directly into our skin—more on this in chapter 5). All these factors combine to sabotage our mental status and make getting healthy feel like walking across an open minefield.

The odds may seem against you, but by the end of this chapter you will be armed with an artillery of your own: the know-how to use food to attain your best health yet, so that you can feel great and manage your weight in the process. I've done my best to sift through and distill all of the latest nutritional information, delivering you only the most potent and pertinent findings to help you stamp out nutrient deficiencies and achieve the body you deserve. And, most critically, you're going to gain a deprivation-free road map to healing your relationship with food.

Avoid Mouth Porn

If there's any one food category that typifies the Standard American Diet, it's processed foods. Born of refined grains, bizarro oils, and stripped of essential nutrients, these foods zap your energy, undermine brain function, and cause you to gain weight. Cutting out this type of food, which includes most breads, pastas, granola bars, sugary snacks and beverages, and cereals, may be the single most effective step you can take for a slimmer waistline.[1]

If you're anything like me, when you crack open a bag of chips or pint of ice cream, or even try the half-decent bread at your favorite restaurant, it takes real mental muscle to stop. Personally, I want to eat until the container or bread basket is empty. If you've been there, you've experienced firsthand the power of *hyperpalatability*. These foods are so pleasurable they are difficult if not impossible to eat in moderation. The good news is, we can predict the effects that these foods have on behavior, because they aren't unique to you. It can be likened to another modern phenomenon: Internet porn.

Presenting an array of camera angles, fantastical fetishes, and professionally maintained bodies, pornography is very exciting to the brain, and its ubiquity today has led some to becoming addicted. To a vulnerable brain, porn can have a drug-like effect, escalating one's desire for an ever-more potent "hit." Porn addicts report an "overriding of satiation mechanisms," a "numbed pleasure response," and an "erosion of willpower."[2] Does that sound familiar? Though porn addiction is a relatively new area for psychiatry, many of the neurobiological underpinnings linking it to food addiction are the same.

What are the ingredients that take a depiction of sensuality and make it *pornographic*, or turn a food hyperpalatable? In porn, it's the combination of forbidden fetishes, willing bodies, and the ease with which it can all be accessed. In food, it's the combination of flavors and textures like sugar, fat, and salt—usually added in by manufacturers with reasonable intent—to create a delicious product that sells. But these essential components that are now freely available hadn't been for most of human evolution. As a result, they are coveted by the human brain for their historical scarcity—and life-saving potential—which persisted until only a short time ago.

FINDING YOUR FOOD REWARD TRIP WIRE

To get a sense of the power of food combining, try this simple experiment at home, which requires only two ingredients: a potato and a stick of salted butter. Bake the potato as you normally would. Keeping it plain, slice it open and, using a fork, scoop up a mouthful and eat it (be careful not to burn your mouth!). You likely won't get very far, as the pure, fat-free starch by itself is not likely prone to overconsumption—or even consumption for that matter. Now combine the salted butter and the potato, letting the butter melt, and taste the combination. What you'll notice is that the fat and salt from the butter suddenly makes the potato delicious. Though you might have predicted this outcome, it illustrates a powerful point: the combination of certain flavors and textures provides a gateway to hyper-palatability and thus insatiable appetite. By being aware of this fact, we may make better choices where our health and weight is concerned both in the supermarket and in our kitchens. Now, don't waste food—enjoy that potato.

Hyperpalatable junk foods now overrun our supermarket aisles, ending up in our shopping carts and, frequently, around our waists. The advice from the food industry (and often parroted by health authorities and fitness gurus alike) is to simply "eat less and move more" to lose weight, but this ignores the reality of how these foods affect our behavior. Moderating your consumption of packaged processed foods is like telling a pornography addict to moderate their viewing of porn, or a drug addict to just use fewer drugs: it doesn't work. In the end, you're left feeling like a failure when you simply *can't* eat less.

In groundbreaking research from the National Institutes of Health, scientists saw firsthand what effect these processed foods have on the consumption habits of regular folks. The researchers took a group of people and, over time, exposed them to either a packaged food supply—think bagels with cream cheese, potato chips, and fruit juice—or perishable, fresh foods. In each phase, subjects could eat whatever they wanted, and their food choices were documented. When eating whole, unprocessed foods like properly raised meats and veggies, subjects ate until they were full and yet experienced effortless weight loss (I'll give you lots of food options in my shopping list in chapter 7).[3] The processed food diet on the other hand was less satiating for the same number of calories. It *created* an average daily energy surplus of 508 calories, mostly from fat and carbs. If sustained, that adds up to an extra pound of fat stored, every week!

Weight gain may be caused by consuming more calories than you burn every day, but hyperpalatable junk foods don't just make it ridiculously easy to overeat. They also influence your daily energy expenditure. That's because the digestion of food itself burns calories, and whole foods burn about double that of processed foods.[4] Combine that with the insatiable hunger that these foods trigger, and say goodbye to your waistline.

Sugar, Sugar, Everywhere

The round-the-clock availability of refined sugar in the human diet is a brand-new phenomenon. Today we overconsume sugar to the tune of sixty-six pounds per person, per year. But in the past, the only sweet food available to a forager would be ripe fruit, and these fruits would only taste a fraction as sweet as they do today. When in season, a hunter-gatherer would likely gorge on what he could lest it spoil or be eaten by other animals; any ambivalence in doing so would mean greater risk of starvation. Our tastes have thus been *engineered* by natural selection, creating a preference for sugar that is likely as deeply embedded as our appreciation for sex—the former to aid in individual survival, the latter to aid in species survival.

Like fat and protein, sugar can take many forms, though for all the choice modern agriculture brings, we are presented only with the *illusion* of variety. Chances are, most of the foods stocked within your supermarket are the processed permutations of just three plants—wheat, corn, and rice. These grains, which look nothing like the sugar your grandmother used to bake cookies with, have just as potent an effect on your blood sugar when pulverized and used to create the bagels, breads and breading, granola bars, muffins, crackers, and commercial cereals we now call "food." *Any* grain flour today, unless you've ground it yourself, should be cause for concern, as it begins to break down and release sugar molecules as soon as it passes your lips, adding to the sugar tsunami in your body's circulatory system.

Here are some "foods" to watch out for. Keep in mind that while they may look different on your plate, their ultimate form is sugar and their final destination is in your blood.

Nothin' Fine About These Refined Grain Products

Bagels	Energy bars
Biscuits	Granola
Bread	Gravy
Buns	Muffins
Cakes	Pancakes
Cereals	Pasta
Chips	Pizza
Cookies	Pretzels
Crackers	Rice
Croissants	Rolls
Cupcakes	Waffles
Doughnuts	Wraps

Whenever sugar is elevated in the blood, regardless of which of the above foods it has come from, a hormone called insulin is released by your pancreas. Insulin shuttles blood sugar into the cells of the liver and muscle, thereby bringing your blood sugar back down to its normal level. The occasional insulin spike can be beneficial, like after a tough workout (see chapter 4). But today, our insulin spikes are frequently stacked one on top of another because of the grain-based sandwiches, wraps, muffins, and snacks that we are bingeing on around the clock . . . to say nothing of the sugar-sweetened beverages and cultivated modern fruits and fruit products we also consume with abandon.

Like they can to any other substance, our cells have the ability to become tolerant to insulin. In an insulin-resistant person, insulin is inefficient at clearing sugar from the blood, causing it to remain elevated for extended bouts of time. There is nothing sweet about this scenario, as sugar is not inert once it's inside your body. Chronically elevated blood sugar creates widespread damage, slowly destroying the body as it inflames and ultimately damages your blood vessels from your toes up to your brain. (The word "flame" in *inflammation* is not a coincidence; the inflammation wrought by sugar can literally scorch your insides—more on this shortly.)

Prior to high blood sugar, however, chronically elevated insulin, or hyperinsulinemia, can create a host of problems. As a growth hormone, insulin creates an environment in our bodies favorable to fat storage. And while the spare tire (or saddlebags, or muffin top) may present aesthetic challenges for some, hyperinsulinemia may manifest visibly in other ways. Male-pattern baldness, skin discoloration, and early wrinkling are all associated with chronically high insulin levels. Perhaps the most tangible sign of chronically high insulin is unsightly skin tags. These benign tumors are the physical manifestations of insulin's powerful growth effect, displayed for all to see on your skin.[5]

LOW-FAT, LOW-CARB, AND THE DIET WARS

There are many roads to the body and health you desire. This is evidenced by the dietary diversity in both world's Blue Zones and by modern hunter-gatherers who live long, are free of chronic disease, and whose bodies would be described as athletic. Some, like the Okinawans, consume low-fat diets centered around fish, rice, and tubers. Others do well with high-fat diets, like the Maasai, who incorporate the meat, milk, and blood from their cattle. The one thing they all have in common? Their foods are whole and unprocessed, save for the limited processing they do by hand. The low-fat diet of the Okinawans, for example, does not contain the depleted fat-free or low-fat processed foods that line our supermarket shelves—they're consuming real foods that are naturally low in fat.

For these reasons, I suspect that average folks navigating the modern food environment will find it easiest to adhere to a plan centered around whole foods that minimize glycemic variability (i.e., blood sugar spikes) and maximize nutrient density. Indulge in fibrous veggies, low-sugar fruits (avocado, citrus, berries), properly raised proteins (fatty fish, grass-fed beef, pastured pork and eggs, and free-range chicken), and Mediterranean fats like extra-virgin olive oil at every meal. Avoid processed foods—this goes for breads, pastas, grain and seed oils, and even purportedly "healthy" products like granola and protein bars. This will help to ensure powerful nutrition for your cells and a more effortless regulation of your hunger, *without* needing to count calories.

Insulin being elevated also means that your ability to burn fat is blocked.[6] The tendency of carbohydrates to spare your fat stores is an act of frugality, meant to preserve your hard-won fat during times of plenty so that you might survive the long winter. But fat burning is a healthy and beneficial process—not just for the body-conscious— and many tissues in your body would quite enjoy using fat for fuel if only given the chance. Today our ability to burn fat has been hijacked by the persistent influx of cheap, easily digestible carbohydrates in our diets.

Your heart muscle, for example, loves to burn fat; it is designed to derive 40 to 70 percent of its energy from fat.[7] The majority of your brain's energy requirements (up to 60 percent) can also be furnished by fat, once fatty acids are converted by the liver into compounds called ketone bodies, or ketones. Ketones are often described as a "super fuel," since they are not just passive energy precursors. When present in circulation (a natural state known as *ketosis*), they help to increase the neurotransmitter GABA, exerting a calming effect on the brain. This is partly how ketogenic diets are thought to treat epilepsy and, research is beginning to show, may also aid memory for those with Alzheimer's disease.

Ketones also are able to increase the expression of other powerful compounds like BDNF, or brain-derived neurotrophic factor. This Miracle-Gro protein has antiaging properties that help to promote the growth of new cells in the brain's vulnerable memory center, called the hippocampus. It's perhaps no surprise that reduced levels of BDNF are associated with diseases like Alzheimer's, but also conditions like depression (more on this in chapter 4). By raising levels of BDNF, ketones may support neuroplasticity, which helps your brain stay resilient as it ages. Finally, ketones are thought to cool off damaging oxidative stress, a feature of a wide array of neurological conditions including Alzheimer's disease, Parkinson's disease, autism, epilepsy, and even aging itself.

FAQ

Q: Should I be in ketosis all the time?

A: Certain neurological conditions may warrant round-the-clock ketosis. However, for the average person, full-time ketosis may be unnecessary and can in fact be suboptimal. When we're in a fasted or ketogenic state, our cells begin to clean house, initiating the process known as *autophagy* whereby old and worn-out proteins, organelles, and even cells get recycled. But fed state physiology is important, too; it is associated with repair, protection, and rebuilding. To be in optimal health, a balance between both must be struck on a daily, weekly, or even seasonal basis. Minimizing your intake of sugar and grains—particularly the refined ones—and combining that with exercise and short fasts, which I'll explain in the next chapter, encourages metabolic flexibility and allows for intermittent ketosis. Don't fear veggies or in-season fruit. Even starches have a place; in chapter 4, I will reveal how to use them to boost your energy levels.

Geneticist Sam Henderson, who studies ketones as a potential treatment for conditions like Alzheimer's disease and whose life's work was also motivated by his mother's illness, has written: "The inhibition of [ketone production] by high-carbohydrate diets may be the most detrimental aspect of modern diets." And yet many of the benefits of ketone production become available when you demote grain-based foods and sugar to occasional indulgences, which will reduce insulin, exhaust your body's sugar stores, and allow it to turn on the fat-burning, ketone-generating machinery.

Build Me Up, Buttercup

Sugar isn't the only historically valuable nutrient. Fat, too, is coveted by the human brain. It's evident in our preference for marbled meats and putting half-and-half in our coffee; fat lends an unmistakable creaminess to foods and allows their flavors to linger on our palates longer.

A few decades ago, where there was fat, there was nutrition. Eggs, nuts, fatty fruits, and the flesh and entrails of wild game and fish provided an abundance of fat-soluble vitamins, minerals, and essential fatty acids like the omega-3s and omega-6s. It's believed that access to these nutrients, combined with the advent of cooking, actually helped supply the raw materials needed to create the modern human brain. As a result, we have developed an eons-long appreciation for natural and delicious fat-containing foods, a fact once again exploited by the modern world to create and sell junk foods to the masses.

A major source of fats for modern humans are animals and animal products, but today's livestock do not provide the same nutritive value as those our ancestors hunted. For one, they're different animals. Take cows, which are the purposefully cultivated descendants of wild oxen and date back only about ten thousand years (we've been anatomically modern for two hundred thousand years). Only when they are raised on a pasture and able to eat their preferred diet of grass are they similar to wild game in terms of nutrient content. The artificial diets fed to the majority of modern livestock not only make them fattier with a higher proportion of saturated fat but also cause them to store fewer of the nutrients we require for good health.

SATURATED FAT: WHAT'S THE DEAL?

While official agencies were busy maligning saturated fats for the past fifty or so years, they had seemingly forgotten one simple, but critical, fact about them: nature had already included saturated fats in every healthy, fat-containing food. Raw nuts, seeds, extra-virgin olive oil, cacao, avocados, and even human breast milk—arguably nature's perfect food—all contain significant amounts of saturated fat. Some saturated fats have even been demonstrated to possess healthful qualities, like stearic acid, which may improve functionality of the mitochondria, the energy-generating power plants of your cells.[8] Thankfully, stearic acid is found in abundance in dark chocolate and the fat of grass-fed beef.

Still, it must be said that a healthier diet will naturally include fewer of these fats, as properly raised animal products like grass-fed beef and wild salmon contain lower proportions of them than their industrially farmed counterparts. There's also preliminary evidence that certain genes, such as the ApoE4 allele, carried by 25 percent of the population, may predispose carriers to cholesterol issues with more saturated fat in the diet.[9] As a general rule, there may be no need to limit saturated fat *when contained in whole, unprocessed foods* like grass-fed beef and wild salmon, but, as always, individual experimentation is key.

Two such casualties of the industrial food complex are the omega-3 fats docosahexaenoic acid (DHA) and eicosapentaenoic acid (EPA), which are normally found in the fat of fatty fish, pastured cows, and the eggs of pastured hens. Since we're eating fewer of these foods, your average person isn't getting enough DHA and EPA fat. Why does this matter? DHA is an important component of healthy cell membranes and allows cells to receive energy and other important molecules. In your brain, this may mean a more balanced mood and a sharper memory; for a muscle cell, faster access to fuel. EPA, which often accompanies DHA, supports the immune and cardiovascular systems, as well as the ability to burn fat and build muscle.

Along with the decline in omega-3s, we're also eating more omega-6 fats than ever before in human history. We need these fats in comparable amounts to omega-3s, but we now overconsume them by at least an order of magnitude, which may come with considerable consequences, as you'll soon see. They are the predominant fat in grain-fed beef, farm-raised fish, and man-made oils like canola, corn, soybean, and mysterious "vegetable" oil. These oils now make up a significant portion of our caloric intake, up from virtually zero percent at the turn of the twentieth century (during that time, the soy industry has enjoyed a 2000 percent increase in the use of soybean oil alone). Here is a list of the oils to avoid:

The Ominous Oils

Canola oil   Safflower oil

Corn oil   Soybean oil

Cottonseed oil   Sunflower oil

Grapeseed oil   "Vegetable" oil

Rice bran oil

The idea that these oils are a heart-healthy alternative to traditional fats like butter, tallow, and even extra-virgin olive oil (which is approximately 15 percent saturated fat) is promoted by billions of dollars of industry money, and my family was one of the countless millions duped. Because they contain no saturated fat, they usually lead to lower cholesterol levels.[10] But by solving the "problem" of high cholesterol, these oils have actually created countless others, and may in fact sit in the driver's seat of many of our modern ills including dementia, cancer, and, yes, even heart disease.

Ever wonder why your grandparents didn't cook with canola oil? That's because prior to a few short decades ago, we didn't have the chemistry labs required to create it. Canola oil, corn oil, soybean oil; all these tasteless, multipurpose oils are the result of harsh industrial processes including the use of caustic chemical solvents. The very nature of production harms the fats, which are no longer protected by the antioxidants found in their whole food form. In doing so, it allows a form of chemical damage called oxidation to occur, and this process progresses through to the storing, shipping, and cooking of these delicate oils. By the time dinner is served, it's *The Walking Dead* on a plate.

Studies have confirmed that we become what we eat; levels of linoleic acid—the type of omega-6 fat found in such oils—have increased in human adult fat cells by 136 percent in the last fifty years alone.[11] But your fat tissue isn't the only place where these oils end up; they easily integrate into particles in your bloodstream called lipoproteins. You may have heard of one lipoprotein, LDL, which is often referred to as your "bad" cholesterol. When tugging along damaged fats, they become unwitting drug mules, creating problems like atherosclerosis and the process known as inflammation. In reality, these particles themselves are good, at least initially.

For a hunter-gatherer, inflammation was a life-saving function of the immune system. It was meant to safeguard us against pathogenic threat (i.e., infectious diseases from "bad" bacteria), or to help promote healing in an injured part of the body, and we still rely on it to keep us healthy today. But inflammation is never a benign state of affairs. It's a fight in a public park, and it leaves collateral damage. When inflammation is temporary, healing can occur; this is how we recover from cuts, scrapes, and bruises or even the occasional infection. Today our immune systems have become chronically activated, not in response to any of the above-mentioned threats, but to what we're eating.

Coursing through our veins, chemically mutated fats rally our bodies' police force—the cells of our immune system—into an action flick–style chase, wreaking havoc everywhere from the lining of our blood vessels to the neurons of the eyes and brain.[12] Over time, the immune response accelerates the aging process while setting the stage for conditions like Alzheimer's and coronary artery disease. The havoc doesn't end there: inflammation can lead to breakage of the strands of genetic material that make up your DNA, the code that operates the life cycle of each cell.[13] This malevolent process (which is also wrought by exposure to radiation and to ultraviolet light, among other things) has been implicated as an early event in the development of tumors.[14] It's perhaps no wonder at least 25 percent of human cancers have been attributed to chronic inflammation.[15]

The good news is we are equipped to repair DNA with the help of enzymes that recognize and correct the damage, but there's a catch. We need the appropriate nutrition in order for these worker chemicals to do their jobs. With 90 percent of Americans now deficient in at least one vitamin or mineral, our bodies don't always have what they need to repair damaged DNA.

MAGNESIUM FOR DNA PROTECTION AND REPAIR

Magnesium is a mineral that we must consume in relatively large quantities for good health. It's in high demand around the body, where it is a cofactor in hundreds of enzymatic processes. One of the key roles it plays is DNA repair; it is required by almost all of the approximately fifty DNA repair enzymes.[16] In a study of almost 2,500 people for whom genetic and dietary data was available, low dietary magnesium intake was shown to be associated with poorer DNA repair ability and an increased risk of lung cancer.[17] Unfortunately, half of the population might not consume adequate magnesium, but luckily it is relatively easy to find. Green leafy vegetables such as spinach and Swiss chard, almonds, seeds, and dark chocolate are among the top sources.

Trans fats are another type of fat that, in their most commonly consumed form, causes direct harm to the brain and body. Their most well-known source is partially hydrogenated oils. These mutant oils are formed when grain and seed oils are chemically altered so that they stay solid at room temperature. For many years, partially hydrogenated oils were used to create smooth and creamy texture in commercial peanut butter, vegan cheese spreads, bakery products, and ice cream. But these fats are about as friendly as a late-winter

Jack Torrance; they are aggressively pro-inflammatory, increasing your risk for early death, heart disease, and worse memory function. New research has even linked higher levels of circulating trans fats to increased risk of dementia, including Alzheimer's disease.[18] Thankfully, hydrogenated oils have been banned by the FDA, but, unfortunately, trans fats still continue to lurk in our food supply.

Grain and seed oils—including canola, corn, soybean, and even mysterious "vegetable" oil—all undergo a process called deodorization, which is a key step in their production that ensures that they are scentless and tasteless. It's the food industry's equivalent of the Witness Protection Program, whereby otherwise bitter, unpalatable oils are made as bland as can be. Manufacturers love it because it allows them to use the same dirt-cheap oil to produce everything from salad dressings to granola bars. Nearly every restaurant uses them to fry and sauté foods, and they're used to make thick and creamy "milks" out of cereals like oats and rice. Here's just a sampling of where to spot them:

Common Grain and Seed Oil Hiding Places

Oil-roasted nuts	Dried fruits
Commercial salad dressings	Cereal products
Mayonnaise	Cereal bars
Anything sautéed or fried in restaurants	Dairy-free milk substitutes
Hot foods and salad bars	Sauces
Grain products	Gravy
Chicken dishes	Pizza
Olive oil "blends"	Pasta dishes

The process that allows all of these foods to incorporate the same tasteless oil—deodorization—*creates* a small but significant amount of trans fats. This fact is nowhere to be found on nutrition labels, which are only required to disclose trans fat content of .5 gram or higher *per serving*. Despite the claim of zero trans fat, "virtually all vegetable oils [and the foods that contain them] contain small amounts of trans fat," wrote Guy Crosby, adjunct associate professor of nutrition at the Harvard T.H. Chan School of Public Health. With your average person now consuming approximately 20 grams of canola or other vegetable oil per day, the amount of trans fats we still consume is significant. There is no safe level of man-made trans fat consumption.

To be sure that your fats are safe and healthy, make sure to consume them as nature intended: in properly raised animals or animal products, nuts, seeds, and fatty fruits like olives and avocados. Extra-virgin olive oil—which humans have been pressing and consuming for eight thousand years—should be the predominant oil used in your home, to both cook with and drizzle raw on top of your food.[19] Your beef should be 100 percent grass-fed, your pork and eggs should be pastured (omega-3-enriched is also okay in the case of eggs), your fish wild, and your chicken free-range. Though these versions can be more expensive, buying in bulk either locally or over the Internet is one way to cut costs. And remember: nothing is as expensive as disease.

FAQ

Q: Max, your food recommendations are great. I just can't afford them!
A: It's commonly assumed that healthy food is more expensive, but with a little planning it can actually be cheaper. One study from Deakin University's Food and Mood Centre found that when people switched from a processed food diet to a whole food–based diet, they were able to save 19 percent on their food costs.[20] Here are some tips:

▸ **Stick to a smaller shopping list of nonnegotiables.** Fresh produce. Eggs. Nuts. Free-range chicken. Grass-fed beef. Extra-virgin
   olive oil. Cut the packaged foods, fancy beverages, branded products, and other nonessentials.

▸ **Cut down on ingredients.** In delicious and healthy Mediterranean-
   style cooking, less is more. Cut the pricey sauces and seasoning
   mixes and grab a high-quality bottle of extra-virgin olive oil and
   basic spices like salt, pepper, and garlic. You'd be surprised at the
   range of delicious dishes you can make with these four staples. In
   the plan section I will offer more simple staples.

▸ **Buy 100 percent grass-fed ground beef and lamb instead of
   steaks.** Ground is always cheaper, and for the money, you can't get
   better in terms of nutrition. And because a cow's diet mainly influences the quality of its fat, if grass-fed is totally off-limits or difficult
   to find, you can still enjoy beef; just stick to leaner cuts or grinds.

▸ **Buy whole poultry.** In the United States, precut chicken breasts
   can cost almost twice the price per pound as a whole chicken or
   turkey. Learn how to break down a chicken so you can store the
   various parts as needed. When in a rush, stick to thighs and drumsticks, which are inexpensive as well.

▸ **Frozen is okay!** Frozen produce is in many ways as nutritious as
   fresh, although some nutrients, like folate, do get lost over time.[21]
   Try to incorporate a mix of fresh and frozen.

▸ **Buy in bulk.** You can shop at wholesale supermarkets or make use
   of the numerous online stores that ship frozen items to your door-
   step, ranging from grass-fed beef and wild fish to fresh produce.

▸ **Know when organic matters and when it doesn't.** No need to
   buy everything organic, just wherever you eat the "skin." For example, avocados and citrus can be conventional, but you should
   always try to buy berries, tomatoes, apples, leafy greens, and bell
   peppers organic (more on this on page 163).

Be Salty (No, Really)

It is perhaps the third ingredient added to packaged, processed foods that is the most delectable to the brain, and the most contentious: salt. Sodium, the primary mineral contained in salt, is required in relatively large quantities for good health. It's fundamental for healthy brain function, and low sodium levels have been linked to cognitive decline among otherwise healthy older people.[22] In fact, severe sodium deficiency can actually mimic dementia, in the form of a treatable condition called hyponatremia. It's also the electrolyte most easily lost in our sweat and urine, making it important to replace if you consume coffee or tend to exercise vigorously (which you should—more on that in chapter 4).

Yet there's a war on salt. The American Heart Association advises to consume no more than 1.5 grams per day for good health. The DASH diet, a government-funded diet to reduce high blood pressure, also advises we cut the salt. And a trip through your average supermarket aisle reveals countless processed foods proudly exclaiming "low sodium!" as if that somehow makes up for the refined grains and unhealthy oils they've chosen to leave in.

Recent and robust research has called into question the notion that salt is bad for us. In a study of ninety-four thousand people, those who had the lowest intake of sodium actually had the *highest* risk of heart disease, with no increase in risk seen at intakes of up to 5 grams per day (and the elevated risk went away when participants simply consumed more potassium, balancing out sodium's effect on blood pressure).[23] Researchers later said in an interview with the Cardiovascular Research Foundation that "about 3 to 5 grams [of sodium] per day appears to be the optimal level associated with lowest risk." (In the chapters to come, I will provide other effective ways of achieving a healthy blood pressure, which is important for better brain function.)

FAQ

Q: What are some top potassium foods?
A: Though bananas get all the notoriety for their potassium content, they are by no means the kings of the potassium kingdom. Avocados, winter squash, sweet potatoes, Brussels sprouts, beets, spinach, and salmon (yes, salmon) also are top sources. For example, a 6-ounce serving of salmon contains one and a half times the potassium of a medium-sized banana, and one whole medium-sized avocado contains roughly twice that.

Truth be told, your average American is already getting enough salt, but that's because most of the salt in the modern diet comes from packaged, processed foods. Even the most innocent-seeming of foods contain massive levels of sodium. If you thought cured meats or salty snacks were the biggest offenders, think again: bread and rolls are the number one source of sodium in the American diet, according to the Centers for Disease Control. These foods (including even healthier options like canned legumes and fish) now account for 75 percent of our daily salt intake, where unnaturally pure extracts of sodium are used to preserve and enhance their flavor.[24]

But it's the *foods*—not simply the salt that they contain—that are the problem, and when cut from your diet, there should be no concern over adding salt to your food. In fact, one of the major benefits of salt is that it can be used to make healthful but admittedly bland foods like broccoli, Brussels sprouts, and squash more palatable for you and your loved ones.

FAQ

Q: What's the healthiest salt to buy?

A: Modern table salts are unnatural distillations of pure sodium chloride (aka NaCl), usually combined with added iodine, a small amount of sugar to stabilize the iodine, and anticaking agents. Many people looking for a more natural alternative have begun shifting away from these processed salts to pure sea salt, though many commercially available sea salts have been found contaminated with microplastics from our increasingly polluted oceans (more on the health dangers of plastic in chapter 5). The best salts, then, are minimally processed and come from pristine sources. Pink Himalayan salt is one option, containing over eighty-four minerals and trace elements, including calcium, magnesium, potassium, copper, and iron. If you go that route, just be sure to attain adequate iodine from your food (some top sources include turkey, shrimp, and sea vegetables like kelp). Visit my Resources page at http://maxl.ug/TGLresources for better salt options.

Prior to the ubiquity of processed foods and even the advent of table salt, getting adequate sodium was a real concern. Animals and vegetables do carry sodium within their tissues, and we obtain it when we eat them, but the concentration is low. As a result, our foraging forebears ingested only a quarter of the sodium we do. It was a valuable commodity even as recently as ancient Rome, where soldiers were allegedly paid in salt, and to this day, to not be worth one's salt is a great insult. If you were ever curious as to where the word *salary* comes from, now you know: *sal*, the original word for salt.

Prioritize Protein

In 1838, Dutch chemist Gerhard Johannes Mulder noticed an abundance of similar, nitrogen-rich compounds found ubiquitously throughout all organisms. He described these primary building blocks (with the help of his colleague Jöns Jacob Berzelius) using the Greek word *proteios*, or "the first quality." This was the fundamental chemical structure we would eventually come to know as *protein*.

Protein serves as the raw building materials for the structure of our bodies as well as countless worker chemicals (hormones and enzymes, for instance) that operate within it. It's needed to build and maintain lean muscle, giving you the strength and robustness to navigate the world. It serves as the backbone to our brain's critically important chemical messengers, neurotransmitters like serotonin and dopamine. And proteins are what are used to ferry various fats and other nutrients around in our bodies, in the form of lipoproteins, like your HDL and LDL cholesterol carriers. Both are essentially fat ferries that rely on protein to function.

Protein is also an important part of our diets, which, along with fat and carbohydrates, serves as a macronutrient (*macro* means large). It's essential; protein is one of the nutrients we absolutely need to get from our food. Luckily, it's easy to find protein in foods like beef, chicken, fish, eggs, and legumes. But how much do we need to get in a day? Our recommended daily allowance (RDA) for protein is .8 gram per kilogram of body weight (or .36 gram per pound of lean body weight, which is your weight minus any extra pounds you may be carrying). Unfortunately, this guideline is meant only to ensure population non-deficiency; it is *not* designed to promote optimal health, and it hasn't changed in over seventy years. Recent research suggests it may be time for an update.[25]

Most adults today consume just over the RDA, leading some to claim we "eat too much protein." But higher protein intakes may actually be ideal according to a recent meta-analysis of randomized controlled trials. The research, published in the *British Journal of Sports Medicine*, found that for weight-training adults across the age spectrum, consuming double the RDA, or 1.6 grams per kilogram of lean mass (.7 gram per pound) daily led to a roughly 10 percent increase in strength and 25 percent increase in muscle mass compared to control groups.[26]

The benefits of having more muscle on the body cannot be overstated. More muscle helps us fight off frailty. It provides a disposal mechanism for excess carbs we may consume. It promotes greater mobility, better balance, stronger bones, higher insulin sensitivity, and lower inflammation. It boosts our confidence and our moods. And it helps our brains avert neurodegeneration. Lest we forget the 3 to 5 percent loss in muscle mass that typically accompanies each decade of life after age thirty, it's important to take care of these precious organs by not skimping on protein, which helps facilitate their growth and maintenance. (In chapter 4 I'll give you a workout plan, too.)

PROTEIN-CONTAINING FOODS

Food	Protein content
Egg (1 whole egg)	6 grams
Chicken breast (6 ounces, cooked)	50 grams
Ground beef, 80% lean (6 ounces, cooked)	42 grams
Wild salmon (6 ounces, cooked)	42 grams
Shrimp, cooked (4 ounces, cooked)	24 grams
Lentils (1/2 cup, cooked)	9 grams
Black beans (1/2 cup, cooked)	7.5 grams

To determine what your daily protein intake might look like, start with your body weight. If you're carrying extra fat, use your goal weight instead. Multiply by 1.6 grams per kilogram or .7 gram per pound (see sidebar below for additional caveats). For example, for a 135-pound relatively lean person, this would add up to a goal of 95 grams of protein per day.

WHEN (AND WHY) TO LIMIT PROTEIN

For healthy people, overconsumption of protein is not something to be worried about because it doesn't often happen. Due to their high satiety, high-protein foods are self-limiting. Think about it; when was the last time you gorged on fish or chicken breast? While high-protein diets have been validated as safe for people with normally functioning kidneys, it may be prudent to limit protein if you have existing kidney disease.[27] In addition, protein increases certain growth factors, so it's always a smart idea to combine the above recommendations with strength training (covered in chapter 4). If you are unable to weight train, or are looking for general health maintenance, aim for a daily protein intake of 1.2 to 1.6 grams per kilogram (or .54 to .7 gram per pound) of lean mass per day.[28]

Aside from helping us grow stronger muscles, prioritizing protein helps us stay trim and healthy by controlling our appetites. When we don't eat enough protein, we tend to eat more carbs and fat. This phenomenon is explained by the protein leverage hypothesis, which posits that we are driven to consume food until we meet our minimum protein requirements.[29] Feeding studies have shown that protein is the most satiating of the macronutrients, suppressing food and even calorie intake more than carbohydrates or fat.[30] Trying to curb hunger? You may want to try protein-dense snacks like full-fat Greek yogurt, unsweetened jerky, or canned fish to watch the protein leverage hypothesis at work.

Protein-containing foods can also help you shed body fat by increasing your metabolic rate. This is known as the *thermic effect of feeding*, or TEF. Since digestion is a relatively labor-intensive process, eating anything is associated with an increase in calorie expenditure. However, each food has a varying capacity to stoke the metabolic fire. Processed foods have the smallest thermic effect, while protein-rich foods have the highest—in fact, 20 to 30 percent of the calories you consume as protein get burned off via digestion alone.[31] This, along with its well-documented hunger-suppressing effect, explains why simply increasing one's protein intake usually leads to a spontaneous weight loss.

FAQ

Q: Doesn't excess protein turn to sugar?
A: The fear of many low-carb or ketogenic enthusiasts is that excess dietary protein turns to sugar. The brain needs *some* sugar; recall that it can fulfill only 60 percent of its energy requirements with ketones. To pick up the slack, a process called gluconeogenesis (literally translated, "new sugar creation") occurs, whereby proteins can be converted to glucose. Without gluconeogenesis, a ketogenic dieter wouldn't survive. Luckily, gluconeogenesis is a *demand*-driven process—it occurs on an as-needed basis, and studies in healthy subjects have shown that high-protein meals contribute little to blood sugar.[32] The benefits of eating more protein—maintenance of lean mass, a boost to your metabolism, and appetite suppression—are significant, and likely outweigh any negatives for most people.

Protein may help protect your sexiest organ of all: your brain. In one study published in the *Journal of Alzheimer's Disease*, adults who consumed the highest levels of protein had the least amyloid beta in their brains and cerebrospinal fluid,[33] which forms the backbone of the plaques that clog up the brain in Alzheimer's disease. The study, which examined almost one thousand cognitively healthy adults, noted a clear inverse dose response—higher protein intake, lower probability of high amyloid burden in the brain. Whether this effect is due to the protein itself, or the fact that higher protein consumption may help us cut down on less healthy and potentially proinflammatory foods, remains a question mark.

When increasing your protein intake, keep in mind that not all protein is created equal. Today meat eaters tend to consume only the muscle tissue of our livestock and toss aside more collagenous parts of the animal (often, these parts are rendered into commercial dog and cat food). Some of these "throwaway" parts contain nutrients that help us metabolize others found in muscle meat. The relationship between glycine and methionine, two amino acids found in protein, perfectly illustrates this.

FAQ

Q: Can't I get all the protein I need from plants?
A: While animal-sourced protein is the among most concentrated and highest-quality protein, it is absolutely possible to get adequate protein on a plant-based diet—but you're going to have to work for it. Prioritize beans and lentils, and incorporate variety. Fermented soy like tempeh can also be an option, but make sure to go organic, as soy is a major commodity crop and subject to heavy spraying of industrial agrichemicals. And keep in mind that some protein-containing foods

like nuts are actually much higher in fat than protein, which can carry a huge caloric load if being used to reach your protein goals. This can inadvertently cause you to pack on the pounds if you aren't careful.

Glycine makes up one third of collagen protein, which is found mostly in animal parts we don't like to eat like skin, connective tissue, and ligaments. It's anti-inflammatory and supports good sleep and detoxification, among other things. Methionine is also essential, but it's concentrated in animals' muscle tissue, which is what we consume nearly exclusively today. We need a balance of both, but most omnivores today tend to consume too much methionine and little glycine, which may have consequences if animal studies are any indication. Rats fed lots of methionine but deprived of glycine have shortened lifespans, and yet live to a ripe old age once small amounts of glycine are added to their diets.[34] And mice on normal diets seem to live significantly longer—by 4 to 6 percent—when given glycine as well.[35]

The more methionine we ingest, the more glycine we need. Take the example of the picky eater who chooses only skinless, boneless chicken breast, which is rich in methionine. That person—by consuming only lean muscle tissue—is raising their needs for glycine, while getting very little of it. Instead, that person could choose to consume all parts of the chicken, which would provide healthy levels of both. (You can always tell when the meat you're eating is rich in glycine-abundant collagen. It's sticky. Collagen is sometimes referred to as the glue that holds the body together, after all.)

How much glycine does an omnivore need for good health? One research calculation estimated that we need to consume roughly 15 grams of glycine daily for our metabolism to function properly, which equates to about 45 grams of collagen.[36] For a "set it and forget it" method to protein consumption that also supports longevity, include collagen-rich joints (chicken drumsticks or thighs, for example), organ meats, and bone broth along with your usual cuts. And eat the skin! If that proves difficult, a basic collagen supplement, which is tasteless and can be mixed into your coffee or tea, can be an inexpensive alternative.

THE ANTIDEPRESSANT POWER OF YOUR FORK

Depression has long been associated with poor diet, but the direction of causality has remained a question mark. Being depressed can cause us to seek comfort with our favorite junk foods, but can those foods actually be fueling the fire that's making us seek them out to begin with?

One of the most encouraging studies was published in 2017 from Deakin University. The study found that, for patients with major depression, cutting out junk foods and focusing instead on fresh vegetables, fruits, raw unsalted nuts, eggs, olive oil, fish, and grass-fed beef improved symptoms by an average of about eleven points on a sixty-point depression scale. By the end of the trial, 32 percent of patients had scores so low that they no longer met the criteria for depression! Meanwhile, people in the group with no dietary modification improved by only about four points and 8 percent achieved remission.

Since that trial, a 2019 meta-analysis of the small but growing body of literature has confirmed that for even nonclinically depressed people, a nutrient-rich diet (especially one that encourages healthy weight) can have a meaningful mood-boosting effect.[37] Combine that with the raw power of exercise—detailed in chapter 4—and its game, set, match!

The Departed Ionome

Have you ever looked at a Nutrition Facts label only to be left wondering how *on earth* you're to get all of the vitamins and minerals you need every single day? I have. The reality is, for even the most health-conscious among us, getting adequate nutrition today can be a full-time job. This is partly because our food has become less nutritious over time. It's the dietary equivalent of the sci-fi mystery series *The Leftovers*, with the difference being it's our nutrients that have departed.

It's no secret that modern agricultural practices place volume as the utmost priority. From fertilization and irrigation to genetic modification, these methods are usually meant to lower cost and increase output. This benefits the bottom line and feeds a hungry population, but the impact that they have on our food's nutrition—its *ionome*—has largely remained a mystery. That is, until Donald Davis, a biochemist at the University of Texas, sought to quantify it.

Dr. Davis compared data on forty-three different fruits and vegetables that was available from the years 1950 to 1999. Over a mere fifty-year period he noted "reliable declines" in numerous nutrients including calcium, phosphorus, iron, riboflavin, and vitamin C.[38] The reductions for some nutrients were substantial; some had declines of up to 38 percent, which was the case for riboflavin. Riboflavin, aka vitamin B_2, is important for healthy cognitive function and mood, and low riboflavin is linked to poor iron utilization, skin problems, oxidative stress, and high blood pressure, among other things.

Our growing methods aren't all that's changed. Carbon dioxide (CO_2) is naturally released into the atmosphere by our oceans, in our breath as we exhale, and via any number of industrial gas and coal-burning processes. It's also a nutrient required by plants to create energy. In the past, our atmosphere would provide plants just enough CO_2 for measured growth and maximal nutrient density.

Since the industrial revolution, atmospheric carbon has nearly doubled to a concentration higher than it has been at any time in the past four hundred thousand years. We know that this has contributed to rising sea levels, but what consequences might it have wrought on our produce?

While it's difficult to measure the total impact, we *are* able to quantify the effect of rising CO_2 levels on food crops thanks to a technique called free-air carbon dioxide enrichment, or FACE. This technique involves blasting large swaths of land with CO_2-enriched air and seeing how the plants grown on that land develop. In the past, FACE had been used on a number of individual crops, but it wasn't until a mathematician and biologist at Bryan College of Health Sciences named Irakli Loladze came along that the net effect rising CO_2 levels have on a wide range of plants was quantified.

In 2014, Dr. Loladze performed a meta-analysis of available FACE data for a number of crops, including spinach, radishes, cucumbers, berries, and various types of rice. He discovered that, over time, elevated CO_2 levels reduced the overall concentration of twenty-five important minerals—including calcium, potassium, zinc, and iron—by 8 percent on average.[39] He also found that the concentration of carbohydrates relative to protein had been increased, further diluting their nutritive value with more starch and sugar. Like humans, plants are now facing their own breed of obesity, which may in turn be contributing to ours!

"Modern crops that grow larger and faster are not necessarily able to acquire nutrients at the same, faster rate, whether by synthesis or by acquisition from the soil," Donald Davis, the University of Texas biochemist, wrote in an op-ed in the trade journal *Food Technology*.[40] The plants on our plates have become more—not less—important as a result. He continued: "Our findings give one more reason to eat more vegetables and fruits, because for nearly all nutrients they remain our most nutrient-dense foods." Here are some tried and true ways of maximizing the nutrient density of your food so as to avert any possible nutrient deficiencies.

Opt for Organic Produce If Possible

It's not just vitamins and minerals that have gone AWOL. Plants, without the ability to fight or flee from predation, produce a myriad of chemicals that they use to ward off rodents, pests, and fungal infections. Many of these compounds, such as polyphenols, have been demonstrated to act in our favor. Organically grown plants, because they can't rely on synthetic pesticides, develop higher levels of these beneficial compounds by about 30 to 40 percent—the equivalent of eating one to two extra servings of fruits and vegetables a day![41] If you can't afford organic, or it is not readily available in your area, don't let that be a barrier to eating more fresh, unprocessed veggies, even if they're conventionally grown. Just remember to rinse them well before consuming.

HOW TO RINSE YOUR PRODUCE

Rinsing your produce with plain water can help to reduce pesticide residues on their surface, but recently published studies have shown that we can make our rinses even more effective by following a simple trick. By simply adding a teaspoon of either salt, vinegar, or baking soda to our rinse water, we can make our rinses up to four times more effective. For the greatest reduction, you may want to soak your fruits or veggies for ten to twenty minutes (the timeframes used by the scientists in the aforementioned studies), but let's be honest: doing so is unlikely to be practical under most circumstances. Therefore, a minute or two is likely to be effective, and if you're on the go, simply hold your produce under running water.[42]

Eat One Large Salad Daily

Want a more youthful brain? Aim to consume a big salad every day, a practice associated with reduced brain aging by up to eleven years.[43] Fill the salad with dark leafy greens like kale, spinach, or arugula, and always include a fat source, like an egg, a piece of fatty fish, or a tablespoon or two of extra-virgin olive oil. This allows for the absorption of carotenoids (plant pigments) like lutein and zeaxanthin, which are found in the greens. While the average adult over the age of fifty consumes fewer than 2 milligrams of these two compounds daily, 6 milligrams per day of combined lutein and zea-xanthin are essential to help prevent age-related macular degenera-tion, and 12 milligrams per day may improve your memory in part by supporting your brain's ability to create energy.[44]

Here are some easy-to-find sources of lutein and zeaxanthin:

TOP VEGGIE SOURCES OF LUTEIN AND ZEAXANTHIN (L+Z)

Food (1 cup, cooked)	Combined L+Z content
Kale	24 milligrams
Spinach	20 milligrams
Swiss chard	19 milligrams
Mustard greens	15 milligrams
Collard greens	12 milligrams
Green peas	4 milligrams
Brussels sprouts	2 milligrams
Sweet corn	2 milligrams
Broccoli	2 milligrams

Remember that for these powerful brain-boosting compounds to be absorbed and thus utilized by your body, a fat source *must* be consumed with them. This is one reason why avocados are a near-perfect brain food—not only are they a low-sugar fruit rich in heart-healthy potassium and fiber, they provide both lutein and zeaxanthin as well as a hefty dose of healthy fat to ensure that those compounds don't go to waste.

Once you've filled your salad with greens, remember that there are no rules. I love adding sunflower seeds (which offer a powerful hit of brain-nurturing vitamin E) and herbs like cilantro. Switch it up so you don't get bored; there are infinite possibilities to ensure that your daily "fatty salad" isn't a boring one!

Screw Moderation; Aim for Consistency

People who adhere to the advice to "eat everything in moderation" tend to eat fewer healthy foods, such as vegetables, and more unhealthy foods, such as grain-fed meats, desserts, and soda.[45] Those with the healthiest diets actually eat a relatively small range of healthy foods. Don't feel guilty or unadventurous for buying your favorite foods on loop, provided those foods include staples like eggs, fatty fish, dark leafy greens, grass-fed beef, cruciferous vegetables, and alliums like garlic and onions. I will provide additional shopping list items in chapter 7.

Incorporate a Range of Animal Products *and* Plants

People have varying and clinically relevant abilities to absorb nutrients from plants or to synthesize essential nutrients from plant-based precursors. Two examples include the plant-based omega-3, alpha linolenic acid (ALA), and beta-carotene. Conversion of each into their usable forms (EPA and DHA fat, and vitamin A, respectively) is significantly influenced by genes, and many people are poor converters. By consuming preformed omega-3s (found in grass-fed beef and fatty fish) and true vitamin A (found in beef liver and also in fatty fish), you procure insurance that these crucial nutrients will be readily utilized by your body.

LIVER AND SHELLFISH: NATURE'S MULTIVITAMINS

Liver, whether beef or chicken, is packed with an enormous concentration of vitamins and minerals, but three in particular make it worth eating alone: vitamin B_{12}, choline, and vitamin A. Each of these nutrients is essential for healthy brain function, and nowhere in the supermarket are they more concentrated—and in a form more easily utilized by your body and brain—than in liver. Vitamin A, for example, might be obtained by consuming orange vegetables, which contain beta-carotene, but people have widely varying abilities to do this. In liver, it's ready for your body to utilize in its plug-and-play form. Look for liver from organic, grass-fed, or pasture-raised animals.

Shellfish is also major brain food, containing huge amounts of vitamin B_{12} and zinc. Assuming you don't have a shellfish allergy, try to incorporate clams, oysters, and crab into your diet. Zinc in particular is critical for processes that support brain function and mental health and is easily absorbed from shellfish. Legumes also contain zinc, but there's a catch; they also carry compounds that inhibit zinc absorption.

Stick to Single-Ingredient Foods

Real foods don't have an ingredients list; they are the ingredients. As you discovered earlier, packaged foods tend to be hyperpalatable. This means that they usually combine salt, fat, sugar, and refined grains to induce insatiable hunger. Making matters worse, these foods are stripped of essential nutrients, which are later added back in, usually in cheap, synthetic form (folic acid—the man-made form of folate—is a perfect example of this, and is often added to refined wheat products). Stick to single-ingredient, seasonal foods and learn simple cooking techniques incorporating a range of herbs, spices, and seasonings to ensure the full breadth of nutrition they offer (I'll give you a few starting pointers in chapter 7).

Bitter Is Better

Remember trying coffee, beer, or wine for the first time? You probably didn't enjoy them very much because of their bitter taste. But many bitter plant compounds, developed to ward off smaller critters, are potent health boosters. Think of the spicy polyphenols in extra-virgin olive oil, which give the oil its anti-inflammatory power, or mouth-parching tannins in coffee, tea, and wine, which may boast neuroprotective and anticancer effects. The digestive tract may have even evolved special receptors just for bitter taste, and activating them might impart benefits including reduced inflammation and better blood sugar management.[46]

Unfortunately, bitter flavors (and the beneficial compounds responsible for them) are being bred out of our produce today, with food growers replacing them with higher and higher concentrations of more palatable sugar and starch (this is in addition to the *unintentional* depletion via conventional farming practices mentioned earlier). Still, you can seek these health-boosting bitter compounds in your supermarket; all you need to do is regularly incorporate into your diet foods like ginger, wild berries, arugula, dandelion greens, citrus peels, turmeric, extra-virgin olive oil, cocoa, tea, and coffee.

We've just covered *what* to eat—a combination of nutrient-dense produce and properly raised animal products at every meal. But *when* you eat may be just as critical, and that's the topic I'm going to cover in the next chapter.

FIELD NOTES

▶ Hyperpalatable junk foods (mouth porn) hijack your brain's reward centers, making it difficult to moderate their consumption, so work to avoid them.

▶ Keeping insulin pulsatile (i.e., by choosing to eat higher-carb foods after workouts) can not only keep you insulin sensitive but promote healthy sugar distribution.

▶ Avoid trans fat–containing grain and seed oils, which promote systemic inflammation and collateral damage to your tissues.

▶ Don't fear salt; salt your veggies to taste.

▶ Higher protein consumption (double the recommended daily intake) supports lean mass growth and maintenance, which plays numerous health-promoting and antiaging roles in the body.

▶ Protein can also help with weight loss by decreasing hunger and increasing calorie expenditure, i.e., the thermic effect of feeding.

▶ Our food is becoming less nutritious over time thanks to agricultural practices that aim to maximize yield and increased CO_2 in the atmosphere. This makes it especially important to base your diet around nutrient-dense foods.

2

TIMING IS EVERYTHING

Brenda Chenowith: I think timing is everything.
Nate Fisher: I think you might be right.

—FROM *SIX FEET UNDER*

For as long as humans have existed, we've held reverence for the natural world. It's ingrained in our most sacred myths; theologians have speculated that the great religions of the world are mere metaphors for elements like the sun and seasons. Long before the advent of artificial lighting, we'd celebrate the presence of the sun and fear its absence. Daylight (or lack of it) has guided our every action, from our sleep and wake cycles to our eating habits and even our mating behavior.

Unfortunately, human industriousness has caused some of our most important relationships to become strained. In the previous chapter, you saw how the advent of agriculture traded the problem of food scarcity for a food supply built on a foundation of cheap and easily obtained convenience foods. This has led to unprecedented rates of obesity and nutrient deficiencies. The following pages explore another of the many relationships that have changed, and that is our relationship with time.

Our nights used to be dark, but we've overruled darkness with artificial lighting. Consider the advancement of mobile technology: less than a decade ago, your average cell phone transmitted two colors and was as bright as a dim bulb. Today our smart phones emit millions of colors and at a brightness that can light up an entire room. This and other factors have caused our world to become the equivalent of a Las Vegas casino, and it has left our bodies and brains lost in time. But the house always wins, and our health has become the ultimate debt paid.

This chapter will serve as your reorientation with time. As you will soon discover, we are evolutionarily primed to live by an ancient, daily rhythm. We will examine the power of light and food, two inputs that our bodies have relied on for millennia to know what time of day it is. Today these variables have become little more than an afterthought, but they are crucial for our bodies to time—and optimize—their processes accordingly. Get ready to wind the clock for better mood, digestion, weight, and a longer, healthier life.

Setting Your Clock

You've got the music in you.

—NEW RADICALS

Sure as the sun rises and sets every day, you likely have a daily routine that you follow. You wake up, rub your eyes, and perhaps visit the bathroom. You head into the kitchen for some water, maybe turn on the coffee maker. At some point you go to work, achieving a few hours of intense focus before your stomach starts to rumble and you eat lunch. If you weren't stuck in the office, you'd likely squeeze in a workout around midmorning or late afternoon. Come the evening, you unwind by going to the cinema, or by conversing with a loved one over dinner. Day in and day out, we carry out certain tasks depending on the time.

In many ways, we run on autopilot, displaying proclivities that have existed for as long as we've been human. For our ancestors, the daylight hours were when they would set up camp, forage, explore, and hunt. At night, they'd seek safety, huddling around the campfire with loved ones to eat, tell stories, and go to sleep. Countless millennia later, we still work during the day and entrance ourselves with stories (movies, theater, books, and TV) at night. More than mere cultural constructs, these behaviors have biological underpinnings that are hardwired in us, honed over eons of life on Earth.

The daily rhythm that our bodies follow is called our *circadian* rhythm, which derives from the Latin words for "about day," or circa diem. Nearly every one of our twenty-three thousand genes is subject to circadian influence. The most potent timekeeper of our twenty-four-hour rhythm? A small area of the brain called the *suprachiasmatic nucleus*, or SCN. Measuring half the size of a chocolate chip and comprising about twenty thousand neurons, the SCN is your brain's central clock. Our bodies are highly influenced by this clock, research is beginning to show, beyond merely waking up and going to sleep.[1]

The SCN is nestled deep within the hypothalamus, a region of the brain that controls some of our most fundamental drives, like hunger, thirst, and the urge to procreate. It serves as our body's master metabolic regulator and thermostat. And it links the brain to the rest of the body through its influence on the hormone-secreting pituitary gland. So ancient, it may have existed in rudimentary form prior to the evolution of the brain itself. All this to say, the hypothalamus—and the SCN that it houses—is elemental to our survival.

The SCN guides our daily inclinations by monitoring the light that comes in through—where else?—our eyes. There are numerous light-sensing proteins in the eye, many of which communicate visual information to the brain, creating the images we see. But a mysterious protein called melanopsin has become a focus of circadian research. Found only in a small number of ocular cells that speak directly to the SCN, melanopsin is not involved in sight, and it's only sensitive to bright blue light, which historically would come only from the sun. It seems to serve just one purpose: setting the body's internal clock.

When activated by bright light, melanopsin sets, or *entrains*, the SCN, anchoring it to the morning and setting off the timer that readies our bodies for the various tasks of the day. This includes releasing various hormones, like cortisol and testosterone, and activating peristalsis, or the movement of contents through the digestive system. (Ever wonder why most people start the day with a bowel movement? Now you know.) And it revs up the metabolic engine, burning through stored fuels to provide all-day strength and allowing a storage buffer for any excess calories you might consume.

NOT SLEEPING LIKE YOU USED TO?

Age-related changes that occur to the eye make us less sensitive to light over time. By the age of forty-five, a person has roughly half the circadian-anchoring light sensitivity as their ten-year-old self.[2] This may explain why older adults today, thrust into the same asynchronous, always-on world, are more prone to sleep problems, and why circadian disruption is so common in age-related conditions like Alzheimer's disease and Parkinson's disease. For a middle-aged person or older, it's going to take more daytime light exposure to get the same beneficial anchoring effect, but doing so may markedly improve your sleep at night.

Fortunately, setting your master clock is as easy as exposing your eyes to bright light early in the day. Research shows that for melanopsin—your light-sensing protein—to entrain your brain's clock, the SCN, about a half hour of 1000 lux of light (approximately the brightness of an overcast day) is needed. Take the time to receive the glorious light of day every morning, either by walking to work or standing near a large window without sunglasses, and know that ambient sunlight is always sufficient to properly entrain your circadian rhythm.

Maintaining Your Clock

Easy as getting enough morning light would seem to be, the average American now spends 93 percent of their life indoors.[3] This not only makes circadian disruption widespread; it's also why *maintaining* its rhythm has become one of the central challenges of modern life. Whether not bright enough during the day, or too bright at night, artificial light now drives our bodies' yearning for balance and routine deep into the abyss, causing us all to live in a perpetual state of jet lag.

HAVING SURGERY? DO IT AFTER NOON

Healing may be the latest bodily process to be linked to our twenty-four-hour clock. In a study of six hundred people who'd had open heart surgery, the risk of a post-surgery cardiac event was halved among patients who had the surgery in the afternoon compared to those who had surgery in the morning. And patients who had afternoon surgery had less damage to their heart tissue. What might explain this striking observation? Surgeons themselves are influenced by circadian rhythms, and reaction times and hand-eye coordination tend to be best in the afternoon. But there was also evidence of hundreds of rhythmically timed genes activated in patients themselves, some potentially playing a role in the higher likelihood of tissue damage in the morning. "As a result, moving heart surgery to the afternoon may help to reduce a person's risk of heart damage after surgery," said David Montaigne, cardiologist and lead author of the study.[4]

When the sun goes down, the SCN typically signals to the nearby pineal gland to begin releasing the sleep-inducing hormone melatonin, almost like a biochemical nightcap. This process was seldom disturbed for our ancestors, where the brightest evening lights came from distant stars, a campfire, or the moon—none of which are bright enough to affect melatonin release. Today light-emitting devices like TVs and smartphones can easily reach a brightness capable of entraining our SCN. This causes our bodies to think that it's daytime when it isn't, and the brain suppresses melatonin as a result.

Below are some common settings and their expected light intensities. Note that the lighting in a supermarket or drugstore can easily reach the intensity required to reset your circadian clock, so if a late-night snack run is in the cards, just know that it may affect your sleep.

TYPICAL LIGHT INTENSITIES

Full moon, clear night sky	25 lux
Dim light	5–50 lux
Living room light	200 lux
Office	500 lux
Gym	750 lux
Supermarket/drugstore	750–1000 lux
Outside, overcast day	1000–10,000 lux
Sunny day, shade	50,000 lux
Bright sun	100,000 lux

Minimum intensity for circadian entrainment = 1000 lux

Though it may be most recognized for helping us wind down, melatonin is not merely a sleep hormone; it plays a key role in the sweeping, curative powers of sleep. Some of its health effects were shown when researchers gave supplemental melatonin to patients with type 2 diabetes. Compared to placebo, three months of supplementation with 10 mg per day of melatonin improved patients' markers of inflammation, oxidative stress, blood sugar control, and other indicators of heart disease risk.[5]

HOW ARTIFICIAL LIGHT CAN MAKE YOU FAT

Melatonin is not just key to better sleep and better health; it may also melt away stubborn body fat. Melatonin increases and activates your body's *brown* fat, a rare and beneficial type of fat that—unlike the white fat that accumulates around our waists and elsewhere—burns calories and secretes powerful hormones that support metabolic health.[6] We already know that obesity is linked to poor sleep and poor health; the neglect of this powerful metabolic pathway may offer another explanation. Allow melatonin to be released in the evening unfettered—this will encourage healthy levels of brown fat and help you sleep better as well. (More on brown fat and other ways of increasing it in the next chapter.)

Melatonin also regulates a process called *autophagy*. It's sort of like biology's version of the KonMari method, whereby old and damaged cellular components get recycled. ("Do these old, worn-out mitochondria spark joy? No? Well, then . . . sayonara!") More than just a mechanism for decluttering, autophagy is vital to cellular health and longevity. Certain conditions can arise when dysfunctional proteins that *should* be cleared away aren't. Alzheimer's disease and Parkinson's disease are two classic examples in which junk proteins aggregate in the brain. And while circadian disruption hasn't been established as a causative force for these two conditions, it's not surprising that Alzheimer's and Parkinson's are each linked with dysfunctional internal clocks.

Alzheimer's disease and Parkinson's disease share another common lineage: damaged DNA, which are either mutations or breakages in the strands of genetic material that make you *you*. Some degree of DNA damage occurs slowly over time, and thankfully melatonin promotes DNA repair.[7] Interfering with this sleep hormone, however, hinders this process. By simultaneously increasing a key source of DNA damage (oxidative stress) and reducing your body's ability to defend itself, circadian disruption not only accelerates aging and brain decay but may promote tumor formation. This might explain why shift workers who make up 20 percent of the global workforce and endure regular circadian disruption seem to have a higher predilection for certain cancers.[8]

Thankfully, our brains produce all the melatonin we need, as long as we don't get in their way. By beginning every day with bright natural light (ideally from the sun) and ending with a reprieve from it, you encourage optimal melatonin release, which helps to counter the processes that drive inflammation, cancer, autoimmunity, heart disease, and neurodegeneration.[9] Here are some other ways to ensure that your circadian rhythm (and natural melatonin release) marches on undeterred:

▶ **Limit afternoon caffeine consumption.** Caffeine affects the brain in the way bright light does.[10] Many of us possess genetic differences that lead to slowed rates of caffeine metabolism, warranting caution for those of us who like to consume it past 2 p.m. Make 4 p.m. your absolute cutoff.

▶ **Wear blue light–blocking glasses before bed.** Late-night work sesh? Wearing amber-colored glasses for two to three hours before bed has been shown to reduce light-induced melatonin suppression by 58 percent.[11] You can visit http://maxl.ug /TGLresources for a suggested pair.

▸ **Use Night Shift or a similar blue-blocking app on your smartphone and devices.** These add warmth to the color on your devices and remove the cooler blue hues. Set them to automatically activate to their warmest setting every sundown.

▸ **Turn down the brightness on your TVs and devices.** Most TVs, computers, and smartphones allow you to alter their brightness. Keep them on their lowest comfortable setting at night.

▸ **Use amber-colored night-lights in your home.** These inexpensive lights can be placed in your bathroom, kitchen, and other frequently visited parts of the house to reduce reliance on brighter lights at night.

▸ **Consume lutein- and zeaxanthin-rich foods.** In a placebo-controlled trial, these two fat-soluble plant pigments were shown to reduce eye strain, fatigue, and headaches due to screen usage.[12] Sleep quality also improved significantly. Foods rich in lutein and zeaxanthin include kale, spinach, avocado, and pastured egg yolks.

▸ **Consume vitamin A–rich foods.** Melanopsin—your master time-setter—is a vitamin A–based protein. Aim to get vitamin A from animal sources first, including liver, salmon, trout, eggs, and mackerel, followed by orange-colored veggies like carrots and sweet potatoes.

▸ **Avoid late-night exercise.** Work out when you can, but if you have a choice, try to avoid hard exercise for two hours before bed. It is stimulating to the nervous system and may advance your circadian rhythm, causing you to be less alert the next day.[13]

▸ **Consume adequate DHA fat.** DHA is an important omega-3 fatty acid found in wild salmon and salmon eggs, omega-3-enriched or pasture-raised eggs, and grass-fed beef. It can also be supplemented with high-quality fish oil or algae oil. DHA-deficient rats had a suppression of melatonin secretion, which normalized with DHA supplementation.[14]

I provide some other tricks to optimize sleep more generally on page 50.

The Kitchen's Closed

The SCN influences all of the organs of the body by way of hormones, long-range messenger chemicals, that it tells nearby glands to release into the blood. But our internal organs each have their own, smaller clocks that function on their own twenty-four-hour timer. Ideally, we'd like all these clocks to be synced up. But as with our master clock, our peripheral clocks get no respect in the modern world. The reason? Near-constant availability of food.

The average person eats all day long, typically consuming three large meals while snacking and drinking sugary, calorie-filled beverages in between. Most of us are digesting and metabolizing food the entire sixteen-hour period that we are awake. Even nutritional authorities advise to keep blood sugar "balanced" by eating many small meals throughout the day, but not only is this pattern of food consumption associated with higher hunger levels and body mass index (a measure of obesity), it may be the bedrock upon which the modern epidemic of type 2 diabetes and associated chronic diseases is built.[15]

When the sun begins to set, the SCN begins to wind down the processes that support daylight-associated activities, including eating and activity. With melatonin now beginning to rise and cortisol, your body's waking hormone, reaching its nadir for the day, metabolism—your energy-generating machinery—slows down. Eating within this window causes asynchrony between your peripheral clocks and the SCN. If this were to happen once in a while, it'd be no big deal. But eating right before bed regularly can lead to weight gain and ill health.

The most immediate effect midnight snacking has is on digestion. As night progresses, the body's "kitchen" closes; fewer digestive juices are produced, and the contractions that push food through the digestive tract (peristalsis) begin to slow down. The slowed transit of a nighttime meal or snack means greater time for that food spent in the small intestine, which can lead to excessive fermentation by the bacteria that reside there. This can have numerous consequences, such as painful gas (gas is not normally produced in the small intestine to any significant degree), bloating, constipation, and even small intestinal bacterial overgrowth, or SIBO.

As the day wanes, our bodies also become less effective at processing carbs and sugar, a phenomenon sometimes referred to as "afternoon diabetes."[16] This shouldn't come as a surprise: the evening hours are meant for repair and release rather than storage and growth, though storage and growth is exactly what carbs and sugar promote via their stimulation of the hormone insulin. This hormone becomes less effective at night, thereby negatively impacting how our bodies handle sugar.

FAQ

Q: Will nighttime eating make me fat?

A: Overall caloric balance will determine your weight, but while calories don't count for more at night, late-night food binges may negatively affect the hormones that regulate hunger and energy expenditure. This could make you eat more calories while burning fewer, thereby encouraging fat gain over time.[17] There are other downsides to late-night eating, too. As the day progresses, we also become less insulin sensitive. This means that eating a carbohydrate-rich meal at night may cause your blood sugar to stay elevated for a longer duration than it would for the same amount of carbs consumed during the day. As I mentioned in chapter 1, chronically elevated blood sugar wreaks havoc on your insides—not least of which the blood vessels that go to your brain. The occasional late-night fridge raid is fine (hey, we all do it), especially if you're sleeping well, de-stressing, and exercising regularly. But let's try not to make a habit of it.

How many late-night meals does it take before metabolic consequences are suffered? Not many. Three days of misalignment can cause insulin resistance, the hallmark of type 2 diabetes, and even a single late-night meal (11 p.m. compared to 6 p.m.) can make us worse at handling glucose the next day. If you're trying to keep daytime hunger and energy levels in check (and avert risk to your brain and cardiovascular system), consider an earlier dinner the night before.

Aside from promoting metabolic mayhem, elevating insulin levels at night—easily achieved with sweet desserts and starchy snacks—might also undermine your body's best attempts at staying youthful. One hormone that becomes sharply elevated at night is growth hormone (GH). In adults, GH has been touted for its antiaging potential, as it helps to build collagen (important for healthy skin and joints) and preserve lean mass. It also supports healthy cognitive function and plays a role in the rejuvenating aspect of good-quality sleep. Unfortunately, GH and insulin have opposing effects, and high insulin levels suppress the release of GH.

To be clear, the research on circadian nutrient timing is new and evolving, but experiments in both animals and humans validate the notion that when you eat might be as important as what you eat. One leader at the center of this research is Satchidananda "Satchin" Panda at the Salk Institute for Biological Sciences in La

Jolla, California. Dr. Panda studies how cellular timekeepers work with one another across the body. Most notably, he was part of the team that discovered the light-sensing melanopsin protein, and his lab (which I had the pleasure of visiting in 2018) is now looking at how the body's numerous clocks work together to influence fat storage, disease, and aging itself.

Similar to us, mice typically eat during a certain half of the day, which makes them a good candidate species for circadian research. Dr. Panda wanted to know what would happen if he disrupted their daily rhythms to simulate what the modern human experiences on a regular basis. He and his team took two groups of mice and gave them the exact same diets, meant to mimic the obesity-promoting Standard American Diet. The only thing that differed between the two groups was *when* the mice had access to that food.

Some mice were given access to the food around the clock, while others only had access during the night (when mice normally come out to forage and feed). The results of the experiment were staggering. While the mice that ate around the clock became obese and unhealthy, the group that was only given access to food within an eight- to twelve-hour evening window ended up slim and healthy. Both groups of mice consumed the same number of calories (and the same mixture of unhealthy fats and sugars), but the mice in the nighttime-only feeding group weighed 28 percent less and had 70 percent less body fat after eighteen weeks. Independent of what they were eating, sticking to their natural nocturnal feeding time protected them against obesity and even gave their health a boost.[18]

Now, humans are not mice, but signs are pointing in a similar direction for people who practice time-restricted eating. A number of small trials have shown that merely eating an earlier dinner can improve blood sugar and blood pressure, independent of weight lost.[19] It may even help fight cancer. One study from Spain involving over four thousand people found eating an earlier dinner (before 9 p.m. or at least two hours before bed) reduced the risk of breast and prostate cancers by 20 percent.[20] A UCSD study yielded equally promising findings, this time for cancer recurrence. The study involved 2,400 women with early-stage breast cancer and found that a nightly fast of fewer than thirteen hours was associated with a 36 percent higher risk of recurrence, compared to thirteen or more hours of food abstention per night.[21] There was also a trend toward increased mortality for the late-night eaters.

While more research is needed to clarify who will benefit most from time-restricted eating, it's encouraging to think that those with limited access to healthy food may be able to improve their health by simply honoring their bodies' innate rhythms. Nonetheless, round-the-clock eating is a new phenomenon for *all* of us, driven by an abundance of food that would have been inconceivable to our hunter-gatherer ancestors. Thankfully, the solution is simple: if you can help it, avoid eating for two to three hours before bed. Of course, water and unsweetened herbal teas like chamomile are fine.

Breaking Bad Breakfasts

The previous section was all about the dangers of late-night eating, which can cause your peripheral clocks to believe that it's daytime when it's not. This can send mixed messages to the SCN, the body's master clock. The consequences of this may be poor digestion, increased fat gain, accelerated aging, and possibly even certain types of cancers. But what is to be said of our morning meal—the most *important* meal of the day, if cereal manufacturers have their say?

Just like it does for the latter half of the day, the SCN has plans for your morning. Circulating melatonin should be minimal upon waking, although if you use an alarm clock to get up it's not unlikely that the hormone, which peaks mid-sleep, is left still elevated. This residual morning melatonin can impair glucose control at a time of day when it should otherwise be at its peak. (Scientists demonstrated this when they gave a morning melatonin supplement to a small group of subjects. Blood glucose shot higher and stayed elevated longer after an oral glucose tolerance test compared to placebo.[22]) For this reason, it's probably wise to wait an hour after rising for your first meal, especially if you use an alarm clock to wake up.

Cortisol is another hormone closely tied to our twenty-four-hour rhythm. Though it is commonly referred to as a stress hormone, it's also the body's "waking up" hormone and promotes energy and alertness. It begins to rise just before you wake up, reaching its peak about forty-five minutes after waking and then beginning a decline that continues throughout the day. To power your morning, cortisol helps break down and make various fuels available. It has this effect on all of the tissues of the body, but partly because of the low insulin environment in the morning, cortisol primarily goes to work on your fat.[23] This creates a fat-burning opportunity that any morning physical activity is going to accelerate.[24]

Unfortunately, when we reach for starchy or sugary foods such as the bowl of oatmeal or glass of orange juice immediately upon waking, this pumps the brakes on fat release, allowing cortisol to continue to exert its tearing-down effects everywhere *but* your fat tissue. The high-cortisol, high-insulin hormonal milieu therefore has the unfortunate effect of helping to redistribute your weight from muscle to fat.

HOW CHRONIC STRESS MAKES YOU "SKINNY FAT"

People who are always stressed out tend to develop apple-shaped torsos. Here's why that happens: chronic stress causes cortisol, a stress hormone, to become chronically elevated. When we're stressed, we also tend to reach for rapidly digesting carbs to soothe ourselves. The excess cortisol combined with elevated insulin from the carbs eats away at our lean mass and causes us to store fat. This is why chronically stressed-out people tend to be "skinny fat." Though the term may make you laugh, it's no laughing matter; the fat that chronic stress causes you to store is primarily the dangerous visceral fat in the belly (creating the apple shape). This inflammatory fat wraps around your vital organs and increases your risk of diabetes, heart disease, and brain shrinkage.

The morning fat-burning window is also why many people often wake up in a mild state of ketosis. Ketone bodies, a by-product of fat metabolism, are a useful fuel source for the brain, which I mentioned on page 15. Early-morning ketone availability may partly explain why we tend to feel a sense of clarity in the morning, and ketone production comes to a grinding halt as soon as breakfast in its most common form (grain cereals, muffins, and toaster pastries, for example) is consumed.

If you enjoy breakfast, feel free to eat it, but know that there's no biological necessity for an early-morning meal. Like late-night eating, breakfast is a modern construct, and its stature as "the most important meal of the day" has been implanted in the modern mind largely by junk food manufacturers. Feel free to set your own schedule and enjoy your first meal of the day an hour or two (or three) after you wake up. This will help maximize the burning of fat among other positive things.

Before moving on: once you've found a routine that works, stick to it. A study published in the journal *Obesity* did find that skipping breakfast led to higher hunger levels and less efficient glucose handling (including higher levels of insulin), but *only* for those who regularly ate breakfast.[25] So while we may not have a biological need for that early-morning meal, we may in fact have a need for meal regularity. Opt for ample protein (eggs are a great example) and fibrous veggies or a large "fatty" salad, which are guaranteed to fill you up and provide all-day energy without the crash. And remember, consistency carves mountains.

Rejuvenation Station

When it comes to slowing down the clock, life extension is indeed possible. The catch? There are two: it involves calorie restriction, *and* it has only been successfully demonstrated in lab animals. Studying longevity in humans is a bit more challenging. We don't sleep in labs, we live a lot longer, and we like to eat. (Correction: we *love* to eat.) So while most of us would happily opt for a 40 percent increase on our life spans like food-deprived lab rats seem to achieve, we need a better route to get there.[26]

Thankfully, longevity researchers have begun to look for calorie restriction *mimetics*—compounds or strategies that can *mimic* the beneficial effects of prolonged calorie restriction but without the misery that goes along with it. Some emerging food-based candidates include *resveratrol*, the antioxidant found in red wine; *fisetin*, found in strawberries and cucumbers; and *curcumin*, found in turmeric. The most promising of all, however, may derive from a practice as old as humanity itself: fasting.

For most animals, periodic fasting occurs naturally as a consequence of an unpredictable food supply. For nearly all of human history, we were no different. Prior to the agricultural revolution, our next meal was always a big question mark. As a result, the bodies of our hunter-gatherer ancestors (and those we inherited) have become adapted to remain strong and resilient in the face of food scarcity. Fast forward to the relative abundance of the post-agricultural world, and nearly every major religion has endorsed fasting as a means of cleansing our spirits and reaching for our higher selves. What they hadn't counted on: we inadvertently have been giving our cells and organs a thorough cleansing as well.

But how do the cells of the body know when we've decided to

fast? Answering that question has been mission critical for longevity researchers. Why? Because if we're able to find the signal that tells our cells "food is scarce," we may be able to activate those signals on demand and reap the myriad cellular benefits that ensue. Plus, we'd be able to do it *without* committing to a lifetime of starving ourselves.

The chief nutrient sensor that our bodies use to assess whether or not we are in a calorie-deprived state is—and it's a mouthful—adenosine monophosphate-activated protein kinase or, simply, AMPK. AMPK senses overall energy (i.e., calorie) availability. You may have heard of adenosine triphosphate, or ATP, the basic energy currency of cells. Under normal circumstances, ATP is able to be generated to meet the demands of our activity. But when ATP can't get replenished fast enough, such as during calorie restriction or high-intensity exercise, AMP builds up in the cell. AMP is an energy-depleted version of ATP, and too much AMP leads to the activation of AMPK.

AMPK sits at the helm of coordinating the body's response to the sudden lack of energy. It promotes increased fat burning, better glucose handling, improved insulin sensitivity, and reduced inflammation. It also *decreases* the liver's synthesis of fats like cholesterol and triglycerides.[27] And, since AMPK's duties include making sure your cells are better prepared for next time, it spurs the creation of healthy, new energy-generating mitochondria (dysfunction of these little power plants is associated with aging and numerous age-related diseases). This is all why activating AMPK is considered a powerful lever for the life-extending properties of calorie restriction.

Other Potential AMPK Activators

Astaxanthin (in krill oil and wild salmon)	Cold exposure
	Curcumin (in turmeric)
Berberine	Extra-virgin olive oil
Coffee	Green tea

Heat (e.g., saunas)

Metformin (a type 2 diabetes drug)

Quercetin (in capers and onions)

Reishi mushroom

Resveratrol

Sulforaphane (in cruciferous veggies)

Vinegar

What might you do to support the activation of AMPK? Calorie restriction, of course. Other than that, high-intensity interval training, which I describe in detail starting on page 122, is a potent AMPK activator, precisely because it creates a temporary state of energy deprivation. And new research suggests that a few hours of daily fasting can also activate this pathway. By simply eating less frequently, we allow AMPK to become active, whereas eating around the clock keeps AMPK perpetually subdued. Avoid food for the first hour or two (or three) after waking up and avoid food for two to three hours before bed. (Conveniently, these are the exact same recommendations as above in regard to your circadian nutrient timing.)

Slowing the Clock

For an animal, not having enough available energy is among the most pressing of concerns, and how the body responds could mean the difference between survival and starvation. So when AMPK is activated, it sends out the Bat Signal to alert other pathways involved in helping you persevere. Today we describe these effects as being "antiaging," but for most of human history, these precautions were indented simply to keep us alive.

One pathway that AMPK stimulates is the FOXO family of proteins. One of them, FOXO3, has been proposed as a longevity protein. It boosts stress resilience (important if you want to live a long time) and may help prevent age-related diseases including cardiovascular disease, type 2 diabetes, cancer, and neurodegenerative diseases. Some lucky people have genes that make their FOXO3 more active, and those people have markedly higher odds of living to one hundred. Genes or not, you can activate FOXO3 just as easily.

For FOXO3 to activate, it needs a signal, and AMPK is just that. While eating around the clock keeps AMPK chronically deactivated, constraining your feeding window to eight to twelve hours every day encourages AMPK—and subsequently FOXO3—to ramp up. (FOXO3 is also sensitive to insulin, which acts like a nutrient sensor for glucose availability and is discussed on pages 12–13. By keeping insulin within a normal healthy range with a lower carbohydrate diet among other things, we allow FOXO3 to come out of its hole.)

Finally, there's mTOR, which may be the most potent antiaging protein of all. mTOR was discovered decades ago while scientists were investigating how a strange bacterial compound discovered on Easter Island seemed to exhibit powerful anticancer effects. It appeared to work by inhibiting a protein in the body involved in cell proliferation, which is increased in cancer. The compound was named rapamycin for Rapa Nui, the Polynesian name for the island on which it was discovered, and its target, the anticancer protein, came to be known as mTOR, or mammalian target of rapamycin.[28]

mTOR promotes storage and growth. As with insulin, this can be highly beneficial when that growth occurs in your muscle tissue, which mTOR helps to achieve. It is also an important player in the formation of synapses—the connections between brain cells—and neuroplasticity, which is your brain's ability to change over time. These processes all require mTOR-regulated growth. But mTOR also has a dark side.

Too much mTOR activity has been linked with autism, seizures, and certain cancers.[29] It can even accelerate aging. When activated, it's the central gatekeeper to the house-cleaning process known as autophagy. Autophagy clears away old and damaged cell components, such as old mitochondria—the energy generators of your

YOUR KEY NUTRIENT SENSORS

Sensor	Role	Effect of fasting	Benefits
Insulin	Responds to carbohydrates, and protein to a lesser degree	↓	Liberates stored fat for use by organs such as the heart, eyes, and muscle. Allows ketones to be generated for use by the brain.
mTOR	Responds to dietary protein and overall energy	↓	Accelerates autophagy; worn-out or damaged proteins, cells, and organelles are recycled.
AMPK	Responds to overall energy availability (fat and carbs) or lack thereof	↑	Increases insulin sensitivity, stimulates creation of new mitochondria, burns stored fat and sugar, activates FOXO3 longevity pathway.

cells—making way for new powerhouses to be created. But by being stuck in an always-on state, this rejuvenation process is blocked. We can see this play out in old mice, whose lives can be extended by up to 60 percent by inhibiting mTOR with rapamycin.[30] Rapamycin is not a free lunch, however, and its chronic use is associated with numerous potential side effects such as insulin resistance, the hallmark of type 2 diabetes. This begs the question: is there a healthier way of inhibiting mTOR?

mTOR is sensitive to two things: dietary protein and energy availability. When protein is abundant and energy is flowing, mTOR is revved up. When energy is lacking or protein is restricted, mTOR is inhibited. By limiting our food consumption to eight hours a day—effectively half the feeding time of your average person—we can easily achieve both and spend more time in an mTOR-inhibited state. And while the story on fasting and longevity is still being written, one proposed method has emerged, with clinical research to back it up.

FAQ

Q: Can I drink coffee during my fasting window?

A: Coffee is fine to drink during your fasting window. If black, or with a little heavy cream or your fat of choice, coffee will not elevate insulin or activate mTOR. In fact, some research suggests that coffee can independently inhibit mTOR while stimulating AMPK, the energy sensor of cells that imparts a number of benefits, like improved fat burning and the creation of healthy new mitochondria. Therefore, coffee should not disturb any of the proposed benefits of fasting, and may in fact enhance them.

The Fasting Mimicking Diet

The raw power of activating AMPK and simultaneously inhibiting mTOR was on display with the results of a fasting protocol devised by scientists at the University of Southern California, led by gerontologist Valter Longo. The research suggests that a periodic very low-calorie diet can not only potentially extend life and health span, but even *treat* conditions like multiple sclerosis and type 1 diabetes. It's known as the fasting mimicking diet.

When it was first tested in mice, Dr. Longo and company witnessed essentially a "resetting" of the immune system. The energy-restricted diet destroyed old and dysfunctional autoimmune cells, which were then re-created in a nonautoimmune state during the refeeding process.[31] The rejuvenation of the immune system mimicked what Dr. Longo calls "an embryonic-like program," causing an increase in healthy new stem cells similar to those seen in development. We don't often get to restart with a clean slate, but that's what fasting seemed to do for these rodents' immune systems.

Moving on to higher-level organisms, the human version of the fasting mimicking diet involved five consecutive days of very low-calorie eating. How low? About half of participants' normal calorie intake. And the calories were specifically intended to come predominantly from veggies and healthy Mediterranean fats like extra-virgin olive oil. It was then repeated monthly, for a total of three months. By the end, the subjects had decreased risk factors and biomarkers for aging, diabetes, and neurodegenerative and cardiovascular disease, without any major adverse effects and with just a few days of calorie restriction per month.*

While this research is promising for anyone who wants to live a long and healthy life, Dr. Longo's discoveries also include striking implications for cancer. In the mouse arm of the study, growth factors (the kind that can fuel tumor growth) were so drastically

---

* The diet was also deliberately low in protein, but it's hard to know whether the benefits seen were due to protein restriction, or calorie restriction in general. Independent of calorie restriction, protein restriction hasn't yet proved beneficial in humans—quite the opposite, actually—and is likely a recipe for weight gain and muscle loss, especially over the long term.

reduced that whole organs actually *shrank* and regenerated during the fasting and refeeding process.[32] For cancer patients, his research also suggests that fasting can help sensitize cancer cells to chemotherapy while minimizing collateral damage to normal cells. It's a rare feat when a single modality can benefit multiple systems with minimal potential for harm—and that's exactly what fasting seemed to do.

WHEN NOT TO FAST

Fasting is an exciting field of research, with staggering implications for cancer and other conditions. Just be mindful that every person is different. Not everyone will want to fast or will even benefit from it. For one, advanced cancer leads to dramatic weight loss, called cachexia. In my mother's final months, the absolute last thing I wanted to do was deprive her of anything—in fact, I regularly visited Veniero's, one of New York's top pastry shops, to pick up her favorite desserts: key lime pie and strawberry shortcake. Always speak with your oncologist about potential dietary interventions.

Finally, anyone pregnant, prone to eating disorders, or with a medical condition should also exercise caution when fasting. Some women who try extended fasting experience hormonal and metabolic disturbances, a phenomenon that has been validated only in animal research so far.[33] As always, your mileage may vary; start slow, and pay close attention to the signals your body is giving you.

The takeaway here is that aside from minding your meal timing, occasional low-calorie dieting may be useful for a long and healthy life. It makes sense from an evolutionary standpoint that our bodies would know what to do once food became scarce, since it's unlikely our ancestors had successful hunts all year round.

We've covered a lot of ground in this chapter, but the overarching recommendation is simple: timing is important. From the inputs that our eyes receive every morning to our snacking tendencies throughout the day, our bodies are rhythmic machines. By honoring this fact, doors of well-being open up to us, untethering us from the poor health plaguing our modern world. Up next, an aspect of the natural world we've all but forsaken, leading to stress, immune disarray, and metabolic mayhem.

FIELD NOTES

- ▶ Honor thy body's circadian rhythm. Get plenty of ambient sunlight during the day and avoid bright light at night.
- ▶ Avoid food for an hour or two (or three) after waking, and, if you can help it, for two to three hours before bed.
- ▶ Following these general guidelines will help sync your body's peripheral clock to your suprachiasmatic nucleus, or SCN.
- ▶ Time-restricted feeding, aka intermittent fasting, is a proposed mimetic for calorie restriction that can have numerous health-promoting and life-extending effects.
- ▶ AMPK, mTOR, and insulin are the body's most important nutrient sensors, and by gaining an understanding of how they operate, we can manipulate them for our benefit.

3

THE VIGOR TRIGGER

You didn't come into this world. You came out of it, like a wave from the ocean. You are not a stranger here.

—ALAN WATTS

The art of healing comes from nature, not from the physician.

—PARACELSUS

I grew up in New York City. Though I enjoyed my urbanite up-bringing, my teenage years brought with them the awareness that the hectic city life probably was not doing my mental health any favors. The scant exposure to nature always made me feel a little disconnected, and the long winter months contributed to a form of gloom called seasonal affective disorder.

I was fortunate enough to have had parents who valued time spent outside of the city. Early on in my childhood, my mom and dad purchased a house on the eastern tip of Long Island in a little town called Remsenburg. This allowed my family and me to spend most weekends out amid the pine trees of Long Island. Yes, I still harbored the requisite amount of teenage angst, but I also noticed that every weekend spent away from the city would do wonders for my mood. And though my own observations were all I needed back then, science is now beginning to validate exposure to nature as a pivotal aspect of holistic health, *including* the health of our minds.

Exposure to the great outdoors is more than mere recreation, it's key to living the Genius Life. Vast and unpredictable, nature allows our brains and bodies to experience the world beyond their comfort zones. Even a quick excursion into nature can boost our immune system, reduce stress, enhance our metabolism, and make us happier and less anxious. It can also help us shed fat, and it may even guard us against the march of time. So while we frequently obsess over what we put in our bodies, *where* we put our bodies may be just as important.

Don't Be a Stiff

What good's a beautiful day / if you can't see the light?
—"SUN," BY ED KOWALCZYK, PERFORMED BY LIVE

Sun, glorious sun. In the previous chapter, you discovered how daylight anchors your body's master clock, setting forth a cascade of processes that support living your best life. But your dependence on the sun doesn't end there. Sunlight plays another critically important role in the way your body functions: it causes your skin to generate vitamin D.

One would assume it is easy to get enough vitamin D, but modern life has led to about 42 percent of the US population being deficient. Some of the reasons for this are obvious, like sunblock overuse and the fact that we now spend 93 percent of our time indoors, according to the Environmental Protection Agency. But there are also some not-so-obvious contributors, such as obesity, aging, and deficiencies in *other* nutrients, like magnesium (more on this later). Any or all of these factors may converge to profoundly impact your health, as vitamin D affects your brain, heart, immune system, and even the rate at which you age.

Once present in your blood, vitamin D acts on receptors in cells throughout your body. These receptors influence the expression of about one thousand genes—an impressive 5 percent of your genome. These genes are involved in nearly every aspect of your health and well-being, from protection against cancer and heart disease to helping your immune system function properly. You wouldn't board an airplane with 5 percent of its engine nonfunctional, and you shouldn't let *your* engine run in such a compromised state, either.

Many receptors for vitamin D reside in your brain, where vitamin D modulates antioxidant levels, helping to detoxify and calm oxidative stress. It reduces the overstimulation of neurons, which happens in both Alzheimer's disease and amyotrophic lateral sclerosis (ALS). And vitamin D may also stimulate immune cells to clean up amyloid, the protein that aggregates to form the plaques associated with Alzheimer's disease.[1] One recent meta-analysis identified low vitamin D as the top environmental risk factor for developing the condition.[2] Having healthy levels of vitamin D means better cognition and slower rates of decline (by two to three times) in healthy people as they age.[3] And, while more research is needed, one small placebo-controlled trial found that patients with Alzheimer's disease who had low levels of vitamin D were seemingly able to halt progression of the disease over twelve months with a mere 800 IU per day of supplementation.[4]

CAN VITAMIN D HELP PREVENT DEPRESSION?

One large study of nearly four thousand adults found that vitamin D deficiency was associated with a 75 percent increased risk of developing depression over four years.[5] While correlation doesn't equal causation,

the vitamin D connection remained robust even after researchers controlled for other relevant factors, including pharmaceutical treatment
for depression, other chronic diseases, and physical activity. Vitamin
D aids in the creation and regulation of certain neurotransmitters, including serotonin, the lack of which has been linked to depression.
Many antidepressant drugs boost levels of serotonin, but these drugs
have side effects, can be difficult to wean off of, and tend to be overprescribed, with recent data suggesting that they are most effective
for severe depression. If you're depressed, perhaps getting more sun
is worth a try.

One way vitamin D may help your brain stay youthful is via its
effect on your cardiovascular system. Wrapping around and through
the brain is a network of blood vessels so small that if lined end to
end would stretch an estimated four hundred miles long. It's a har-
bor that facilitates transport of nutrients into (and waste products
out of) the brain, and is a site of dysfunction early on in the process
of cognitive decline.[6] Thankfully, you can care for these blood vessels in many ways: regularly eating nutrient-dense whole foods and
avoiding pro-inflammatory grain and seed oils like canola, corn,
and soy as laid out in chapter 1, and exercising routinely. But vitamin D, synthesized in our skin via the sun's rays, plays a role here
as well.[7]

The arteries that carry blood and nutrients around your body and
up to your brain are meant to be elastic, not stiff. This allows them
to dilate and constrict according to the needs of your ever-changing
environment. Stiff blood vessels can be disastrous for your health.
Arterial stiffness can put you at risk for not only heart disease and
early death but reduced brain volume and cognitive performance as
well.[8] It can also lead to reduced blood flow to the brain, and this

may be especially true for people with increased genetic risk for Alzheimer's disease (i.e., carriers of ApoE4 allele).[9]

Vitamin D, whether from the sun or a supplement, is thought to fight arterial stiffness on two key fronts, especially if one is deficient: it reduces high blood pressure and can dampen chronic inflammation. Here in the West, chronic inflammation and high blood pressure are exceedingly common, but these are not natural or inevitable aspects of aging.[10] People with little exposure to modernization tend to have lower blood pressure and inflammation and more flexible arteries.[11] While myriad variables can explain these differences, don't discount time spent in the sun. In the industrialized world, people with lower levels of vitamin D tend to also have increased arterial stiffness.[12]

WHAT IS THE OPTIMAL DOSE OF D?

To answer the question of optimal dose we must first define an optimal blood range, and unfortunately there is no universally accepted optimal range. In 2014, a robust meta-analysis of thirty-two studies found the lowest risk of early death by any cause (including cancer and heart disease) was achieved with 25(OH)D levels between 40 and 60 ng/mL, and reaching 50 ng/mL might provide cognitive benefits as well.[13] When it comes to dosing, between 2,000 to 5,000 IU of vitamin D per day should be effective for most people to land somewhere within this range, and always choose vitamin D_3 (as opposed to D_2), which is chemically identical to what we create in our skin.[14] If you are overweight or obese, it's possible that you may require a higher dose (more on this soon).[15] Just remember: in biology, you can *always* have too much of a good thing. Too much vitamin D can cause calcium to build up in the body. If supplementing, be sure to get levels checked regularly by your doctor, which can be done via a simple blood test.

If deficient, vitamin D supplements may help to promote healthy blood vessel function, including arterial flexibility.[16] But reducing the benefits of sun exposure down to a supplement may be an exercise in futility. Here's why: the sun's UVA, non-vitamin-D-producing rays, may be beneficial, too, by helping create nitric oxide, a gas that allows our blood vessels to expand and maintain healthy blood pressure. A small University of Edinburgh study found that volunteers who were exposed to the equivalent of thirty minutes of summer sun had a measurable increase in nitric oxide, and this coincided with a reduction in their blood pressure.[17] High blood pressure is often associated with heart disease, which claims more lives than skin cancer every year by a hundredfold. In other words, the sun just may save your life.

Fighting Autoimmunity with D

Type 2 diabetes, cancer, Alzheimer's disease, and heart disease; these conditions were rare in antiquity and now afflict hundreds of millions of people around the world (and make up the leading causes of death globally, according to the World Health Organization). While they are all *multifactorial* conditions, meaning unlikely to have a singular origin, our collective vitamin D deficiency may have a role to play via its effect on inflammation.

In chapter 1, you discovered how food can promote (or reduce) inflammation. Recall that the collateral damage of inflammation knows no bounds—over time, it can scorch your blood vessels and damage your DNA. But what we eat is just one part of the story; the same cells that initiate and carry out the inflammatory cascade all contain receptors for vitamin D. While the role of the vitamin D receptor is still being elucidated, vitamin D deficiency may allow the immune system to run amok like an angry Hulk.[18] Keeping your vitamin D levels within a healthy range may therefore reduce your risk for these inflammatory conditions.[19]

NUTRITION IN THE NEWS

Nutrition science can be confusing, and the media, however well intentioned, seldom helps the cause. This is why it's important to look beyond headlines or media coverage for nutritional truth. One of my favorite examples of confusion about nutritional information comes from a major media site that published an article with this headline: "Millions of Americans Take Vitamin D. Most Should Just Stop." The article was reporting on a robust randomized controlled trial published in the *New England Journal of Medicine* that found that vitamin D did not lower the incidence of invasive cancer. But cancer develops over many years, and after researchers excluded the first two years of follow-up, they saw that vitamin D supplementation *did* result in a 25 percent risk reduction of death from cancer.[20] Cancer can also be "fed" by what drives obesity; cancers associated with being overweight and obese—including pancreatic, breast, and colon—make up 40 percent of cancers diagnosed in the United States, according to the CDC. We're far from figuring out what causes each cancer and for every person, but in this trial participants were overweight on average. This drives home another important point: you can't out-supplement a poor diet and lifestyle!

Another type of immune system derangement has grown exceedingly common: autoimmunity. For the millions suffering globally from autoimmune conditions such as multiple sclerosis (MS), inflammatory bowel disease, and rheumatoid arthritis, the immune system becomes so disgruntled that it actually attacks its own host—talk about biting the hand that feeds! Nobody knows exactly why autoimmunity develops, but seeing as how these types of conditions are increasing in Westernized societies and remain rare in hunter-gatherer societies, the finger points yet again to our modern habitat.[21]

Recent ideas about the origins of autoimmunity have centered around our immune systems' lack of interaction with soil-based bacteria early in life. These forgotten germs, often referred to as "old friends," may be required for development of healthy immune function. But another aspect of nature has also garnered significant interest: our lack of healthy sun exposure.[22] Many who suffer from autoimmunity have low vitamin D levels, and while low vitamin D may be caused by the same factors that lead to autoimmunity, others believe the inverse to be true: that low vitamin D contributes to immune self-sabotage—beginning possibly even before birth.

Multiple sclerosis (MS) is one of the more common autoimmune conditions and has been strongly linked to low vitamin D. In MS, the immune system attacks the fatty sheaths surrounding brain cells, causing fatigue and disability. In the northern hemisphere, children born after the summer have lower risk of developing MS in adulthood than children born after the winter.[23] Mothers pass their vitamin D onto their offspring through blood and milk, suggesting that low vitamin D can precede the disease by decades. (For sufferers of MS, vitamin D also happens to be the only vitamin with sufficient evidence to support routine supplementation, as recently published in *JAMA Neurology*.[24])

One way vitamin D may help prevent or even treat autoimmunity is by increasing a type of immune cell called regulatory T cells.[25] "Tregs" have become a focal point for researchers trying to understand why autoimmunity develops, in part because they serve on the team of immune first responders that help determine whether something in the body is a foreign invader or simply an injured part of itself. Most of the time, Tregs ensure a healthy and appropriate inflammatory response by suppressing the responses of other immune cells, and this includes those that promote inflammation and autoimmunity.

While more research needs to be done before we can make any definitive statements about vitamin D and autoimmunity, there is promise that for some conditions, it may offer a means of treatment, or at least a slowing of progression. For Crohn's disease sufferers, where the immune system inflames the lining of the digestive tract, 2000 IU of vitamin D daily led to remission and an improvement in quality of life for a significant number of patients compared to placebo.[26] Other conditions for which vitamin D may make a helpful impact include type 1 diabetes, lupus, and rheumatoid arthritis. Even vitiligo, a stigmatizing condition in which pigment-producing skin cells are destroyed by the immune system, may cede to vitamin D. One small, open label study found that high-dose vitamin D treatment for six months led to not only a halting of the destruction but a 25 to 75 percent repigmentation.[27]

Seeing as how vitamin D is well-tolerated by nearly everybody, it's certainly worth the effort to make sure yours is within a healthy range. As a bonus, it might even slow the rate at which you age.

Inflammaging

No one knows exactly what drives aging, but excessive inflammation is currently leading the theory pack. The word *inflammaging* has even been used by researchers to describe the close relationship between inflammation and aging. One observational study, for example, found that low inflammation was the chief predictor of good cognition, independence, and long life among Japanese centenarians and semi-supercentenarians (people who have reached the age of 105).[28] But many variables are linked to lower inflammation: deep social ties, unprocessed diets, daily physical activity, and a sense of life purpose, to name a few.[29] What evidence is there that *vitamin D* may serve as a weapon in our arsenal against the march of time?

In one telling study, British and American researchers looked at vitamin D levels and inflammatory markers of 2,160 female twins and found that those twins with the lowest levels of vitamin D had higher levels of inflammation.[30] The researchers also noticed a marked difference in structures called telomeres between the twins. Telomeres have gained notoriety as one of the few proposed biomarkers for aging. They are located at the ends of your chromosomes where they function like shoelace caps, helping to protect your chromosomes from damage (where shoelace caps fray with time, telomeres get shorter). The twins with the lowest levels of vitamin D had shorter telomeres equivalent to five years of accelerated aging compared to those with the highest levels. In other words, low vitamin D was associated with advanced aging, even in age-matched adults with the same DNA.

Admittedly, it's difficult to study aging in humans. The number of variables that convene across the average human life span make ascribing causality to any one impossible. The twin study above doesn't *prove* that vitamin D slows aging—the twins with higher vitamin D levels could have also been more outdoorsy, which can suggest higher levels of physical activity. Unfortunately, most of the data that we have on successful aging in humans comes from observational evidence like the above, *not* experimentation. If a scientist wanted to see how isolated variables like vitamin D affect the life span of an organism, she'd need to find a creature with a short enough life span and reasonable enough similarity to humans to do it.

The nematode is such a creature. It grows to one millimeter long, lives about two weeks, and is see-through. Not the gorgeous stand-in you were hoping for? Surprisingly, the nematode (aka *C. elegans*) has enough in common with us to make it a perfect specimen for aging research: it shares many longevity-related processes and genes, and it also synthesizes vitamin D. Researchers at the Buck Institute wanted to see what the sunshine vitamin might do for an aging nematode, and what they found was staggering. Those nematodes fed vitamin D throughout their adulthoods had an average life span extension of 33 percent.[31] Vitamin D also activated beneficial stress response genes and promoted the maintenance of bodily protein—important considering many age-related diseases are protein-related (more on this shortly).

YOU NEED THESE NUTRIENTS FOR HEALTHY VITAMIN D

The vitamin D that our skin makes must be converted by the liver to 25(OH)D_3, which is what is usually measured when you get your vitamin D levels tested by a doctor. Then it's converted by the kidneys to calcitriol, which is the active hormone form of the vitamin. Unbeknownst to many, the enzymes that perform both of these conversions rely on magnesium. Unfortunately, 50 percent of Americans do not consume adequate magnesium, which could lead to vitamin D staying stored and inactive for a huge proportion of people. Thankfully, magnesium is found in foods like dark leafy greens, almonds, pumpkin seeds, full-fat dairy yogurt, and dark chocolate.

Vitamin K_2 is another essential nutrient that has become depleted from the modern food supply. Found mainly in the fat of grass-fed beef and dairy, and in a Japanese fermented soybean product called *natto*, vitamin K_2 helps guide calcium deposition in the body. This is important, as vitamin D increases calcium absorption from our food. Vitamin K_2 helps keep calcium in places we want it, such as bones and teeth, and out of places we don't, like our arteries, kidneys, and other soft tissue.

Getting Your D

In the previous chapter, we discussed the natural twenty-four-hour cycle. But the sun's daily cycle isn't the only rhythm we've adapted to;

we've been honed by an annual rhythm as well. During the summer months, when the sun's UVB rays are easily able to reach your skin, vitamin D production is easy. But summer doesn't last forever, and depending on your latitude, winter may mean months without direct access to vitamin D–producing rays. How then did our ancestors make it through the dark winter months without risking severe deficiency?

Thankfully, vitamin D gets stored in our bodies' fat tissue to protect us against natural seasonal variability. This also occurs with the other fat-soluble vitamins, A, E, and K. But similar to how storing fat was key to the survival of an early human and today facilitates the obesity crisis, our ability to store vitamin D has become a double-edged sword. Vitamin D can get sequestered by fat tissue, making those who are overweight risk deficiency *even with regular sun exposure*.[32] This also means that if you are overweight and supplementing with vitamin D, you may need two to three times what a lean person requires to achieve healthy vitamin D levels.[33]

Skin color also matters when it comes to vitamin D production. Those with darker complexions have more melanin in their skin. Melanin is nature's sunscreen, and while you may enjoy reduced skin aging, it also makes you more prone to vitamin D insufficiency (deficiency rates shoot up to 82 percent for African Americans and 70 percent for Hispanics, almost *double* the national average). In the summer, ten minutes of exposure for someone with fair skin may be sufficient, but someone with a darker complexion may need up to two hours.

FAQ

Q: Should I wear sunblock?

A: A slew of meta-analyses performed over the past few decades have questioned the effect of sunscreen against melanoma, the most dangerous type of skin cancer.[34] Plus, most drugstore chemical-based sunscreens absorb into the bloodstream at levels that are likely unsafe (more on this on page 156). Still, that doesn't make it smart to burn. Be smart about your sun exposure, and if need be, use a safe, mineral-based sunscreen (e.g., zinc oxide) to prevent sun damage. Remember: in biology, too much of a good thing can become a bad thing!

Our body's ability to create vitamin D also diminishes with age, to the degree that a seventy-seven-year-old person's skin creates half the vitamin D of an eighteen-year-old's skin given the same amount of time in the sun.[35] On top of that, our kidneys, which normally "activate" vitamin D, falter over time in their ability to do so.[36] The takeaway here is that as you age, your needs for sun exposure (or supplementation) increase.

Below is a continuum for how much time you might need in the sun to achieve adequate vitamin D levels. Just be mindful of burning, which leads to unnecessary DNA damage. Ain't nobody got time for that!

SUN EXPOSURE FOR HEALTHY VITAMIN D

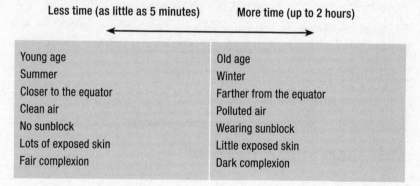

Less time (as little as 5 minutes)   More time (up to 2 hours)

Young age	Old age
Summer	Winter
Closer to the equator	Farther from the equator
Clean air	Polluted air
No sunblock	Wearing sunblock
Lots of exposed skin	Little exposed skin
Fair complexion	Dark complexion

The sun is the ideal way to boost vitamin D because its light, beyond supporting the natural pathways that create vitamin D, provides other benefits like nitric oxide production and setting your internal clock. There's also no risk of "oversupplementing" since your skin will synthesize as much vitamin D as is needed and degrade any extra.[37] Nevertheless, if you choose to supplement, opt for vitamin D_3, which is identical to what we create in our skin, and remember: no supplement will ever fix a poor diet or lifestyle! Revisit the sidebar on page 73 for specific dosing recommendations.

Good Stress

On page 37, you discovered one reason why organically grown produce may be more beneficial to human health. When you minimize a plant's burden of stress with synthetic herbicides and pesticides, you get a plant with less vigor, and this equates to fewer of certain beneficial chemicals present in its makeup. In a similar fashion, humans need stress to become strong and resilient, and nowhere are these stresses more abundant than in the natural world.

You're probably already aware of the body's adaptive response to one type of stress—physical exercise. It makes you stronger. But more than just a means of maintaining a healthy body weight, the strength you achieve in the gym may spill over into unexpected areas of your life, including mental health. It's perhaps why regular exercise is associated with greater tolerance to *psychologically* stressful stimuli.[38] This effect, whereby one type of stress protects you against another, has come to be called *cross-adaptation*.

WHEN GOOD STRESS TURNS BAD

When under any type of stress—be it the stress from work or a workout—your body is thrust out of balance, and the collective steps it must take to recover is known as its *allostatic load*. But pile on the stress and suddenly your load becomes overload. How do you find *your* tipping point? Think of an empty glass; that is your total tolerance to stress. Under normal circumstances, you want the glass to be empty, so that you can add in things like thermal stress, high-intensity exercise, or even caffeine, which can stimulate the body's stress pathways. But if you are already enduring a lot of stress—and not allowing yourself adequate time to recover—your cup may be half full, causing it to overflow with any additional (and potentially beneficial) stressors. Allostatic *overload* not only leaves you feeling crappy and burnt out, it also makes you vulnerable to infection and disease.[39] Always be mindful of your total stress burden, and eliminate the sources of chronic stress first so that your baseline is an empty glass as opposed to one that is already full. This way, you can continue to enjoy exercise, heat or cold stress, and, yes, coffee. On page 124, I dive into recovery and relaxation, both of which can help you reduce allostatic load.

Compared to exercise, its benefits may be less immediately noticeable in the mirror, but your body adapts to *thermal* stress similarly. Hot or cold, our body's response to variations in temperature occur with one objective in mind: to not die. That's because the body is designed to operate at a certain temperature, around 98.6 degrees. When challenged, a cascade of powerful adaptive changes are summoned, which the following pages will home in on as a gateway to greater health, a better mood, and perhaps even the body you've always desired.

Ice, Ice, Baby

Prior to the relative safety of the modern world, dramatic swings in temperature frequently implied physical threat. Imagine, as a hunter-gatherer, fishing for your family on a frozen lake. One day, you pass over thin ice, suddenly falling through and into the frigid water. In seconds, you went from enjoying a routine day on the ice to facing possible—if not probable—death. Your body launches into action: your muscles contract, grabbing on to what's left of the ice and vaulting you out of the water with seemingly superhuman strength.

Much as you'd like to credit your brawn for your survival abilities, your brain played just as crucial a role. That night as you speak of your heroics, you mention feeling as though your experience occurred in slow motion, even though it all took place within an instant. This is common during stressful events, as senses heighten and reaction time quickens. Your brain also takes precautions to prevent it from happening again, logging the minutiae of the event with the crystal-clear precision of a high-speed camera. You share details that on most days you'd have barely noticed; location, the sound the ice made as it cracked, time of day, weather, and the like are all duly noted with crystalline accuracy.

Many of these cognitive effects can be traced to a chemical messenger in the brain called norepinephrine, which spikes sharply during a stressful event. It's known for supporting laser-like focus, attention, and detailed memory storage, and low levels are related to ADHD, feelings of lethargy, and lack of focus and concentration. But the neurotransmitter also plays a role in depression; many antidepressant drugs aim to work by boosting it. Harnessing the power of norepinephrine may therefore provide a lever to both increased mental vigilance and a brighter mood—even if it means getting out of your comfort zone and enduring the occasional cold.

NOREPINEPHRINE AND ALZHEIMER'S DISEASE

The locus coeruleus—the hub of norepinephrine release in the brain—has been highlighted as a potential "ground zero" for Alzheimer's disease, a devastating memory disorder that affects half of people over the age of eighty-five. In the condition, nearly 70 percent of the norepinephrine-producing cells of the locus coeruleus are lost, and the decline of norepinephrine correlates tightly with the progression of the dementia. Sufferers experience a crippling loss of functionality, which is perhaps not surprising given norepinephrine's role in focus, attention, and memory storage, but rodent studies have revealed that norepinephrine also possesses anti-inflammatory abilities and may help the brain better clear the toxic proteins that aggregate and form the plaques associated with Alzheimer's disease.[40] Whether norepinephrine alone can prevent Alzheimer's disease is unknown, but it's perhaps not a coincidence that activities known to boost norepinephrine (exercise, for example, or the use of saunas, which will be discussed shortly) have also been linked with having a protective effect against the disease.

To test this theory with an experiment, Polish researchers administered cryotherapy to a group of patients who were all being treated for mood and anxiety disorders. Cryotherapy (or simply "cryo") involves standing in a gas-cooled chamber for about two to three minutes at a time, and patients were prescribed a regimen of three weeks of daily weekday exposure. After thorough examination, one third of cryo participants had experienced a decrease of depressive symptoms by at least 50 percent, compared to 3 percent of controls receiving standard care.[41] Nearly half of the cryo group also experienced a reduction in anxiety by at least 50 percent; the controls saw no improvement in their anxiety.

Cryotherapy is not without risk, and though the treatment is gaining popularity in larger cities, it can still be pricey. Thankfully, we may be able to exploit the cold for better mood and brain power without the risk of hurting ourselves or going broke. Numerous anecdotes, including one academic case study (published in the *British Medical Journal Case Reports*), have surfaced of people self-treating even severe depression with ice baths, open water swimming, and, yes, cold showers.[42] Even at milder temperatures, brain chemistry changes markedly. One study found that men immersed up to their necks in 57°F water—chilly, but far from freezing—had increased levels of norepinephrine by more than fivefold after an hour.[43] The cold also reduced levels of cortisol, a hormone associated with stress.

To integrate cold immersion into your day, start in the shower. Gradually try lowering the temperature, beginning with about fifteen seconds and extending the time. Just be aware that cold shock can cause an increase in heartbeat and rapid breathing—it activates a stress response, after all. Therefore, it should be approached cautiously. On the other hand, there are no known long-term side effects or withdrawal symptoms if you choose to use cold water therapy to fight depression or enhance your mood—a claim not many pharmaceuticals can make!

NERD ALERT

I'm lucky to have a terrace in my apartment in New York, and in the winter I'll often stand outside for a few minutes shirtless—or even in my underwear—to make use of the "free cryo." It markedly boosts mood and mental vigilance, and it helps me stay lean by activating my body's calorie-burning brown fat, which you'll learn about next. The only downside? The strange looks I get from neighbors and my cat.

Our Internal, Fat-Burning Furnace

Impressive as the benefits of intermittent cold exposure may be, they aren't limited to the mind. Similar to how your wall thermostat kicks the furnace on when room temperature falls below a certain level, your body has its own furnace that activates under similar conditions. The furnace in your body constitutes a group of specialized fat cells that gather around your neck, collarbone, armpits, and spine. These cells comprise your *brown fat*. Brown fat functions a bit differently than the run-of-the-mill white fat cells that we accumulate on our waists and hips. Unlike those fat cells, brown fat cells are packed with mitochondria—cellular energy generators—and they burn calories to keep us warm like an internal heating pad.

In humans, exposure to even mildly cold temperatures spurs brown fat cells into calorie-burning action, a process called *nonshivering thermogenesis* (thermogenesis means "heat creation"). Brown fat is so keen on generating heat that nonshivering thermogenesis can account for up to 30 percent of your metabolic rate.[44] But drop the temperature further, and the body goes into overdrive, torching calories so that your internal organs don't freeze. One human trial showed that immersion in water that was 68°F nearly doubled the subjects' metabolic rate, while immersion at 57°F increased metabolic rate by over threefold.[45] While these increases in energy expenditure may be temporary, other positive health effects are enduring.

In one experiment, researchers subjected patients with type 2 diabetes to regular, ambient cold. Type 2 diabetes is defined by insulin insensitivity leading to chronically elevated blood sugar; an improvement in insulin sensitivity is therefore an improvement in the disease. Wearing only shorts and T-shirts, the subjects spent six hours per day in a room set at 60°F for ten days. Without making any other changes to their diets or lifestyles, patients' insulin sensitivity increased by an astonishing 40 percent—an improvement as good as what could be expected with long-term exercise![46] This makes brown fat a powerful weapon in the fight against not only obesity but aging and degenerative diseases as well.

TAKE A NATURE PILL TO BEAT STRESS

Think you're busy? In Japan, office workers routinely work long hours—*way* longer than their counterparts in the United States. Stress runs rampant, and there's even a Japanese word for being worked to death—*karoshi*—a preventable tragedy that claims a small but growing number of lives every year. Perhaps for this reason (and the fact that 93 percent of the population lives in cities), Japan has become an epicenter for a particular form of therapy called forest bathing. Elsewhere in the world, scientists have caught on that even a quick trip into nature does wonders for our mental health.

One way that nature is able to keep us mentally strong is via its influence on an area of the brain called the subgenual prefrontal cortex. This tiny region is thought to process sadness, guilt, remorse, and negative self-talk. After subjects spent ninety minutes in a natural environment, not only had this area become markedly subdued (evident on brain scans), but the participants showed less rumination compared to controls. People who ruminate more are less forgiving of themselves, and excessive rumination can often forewarn depression and even suicidal ideation.[47]

Other mechanisms may explain why nature is such a potent salve for the soul. It provides an opportunity to bask in the bright light of the sun, which stimulates the production of vitamin D and anchors our bodies' twenty-four-hour cycle. We are cooled (or warmed) by the natural environment, triggering our bodies' ancient thermoregulatory systems along with a number of beneficial brain changes. And we breathe in the scent of nature itself, carried through the air by various plant chemicals that may boost immunity and even the brain's neuroprotective growth factor, BDNF.[48]

At what "dose" do we begin to reap the benefits of nature? Published in *Frontiers in Psychology*, one study found that immersion into nature significantly reduced levels of cortisol, the hormone associated with chronic stress, after only twenty minutes.[49] Whether you go for the "minimal effective dose" suggested by this study, or opt for a full weekend outdoors, a little nature goes a long way. Don't hold yourself captive; get out there and bathe in nature!

Brown or white, our fat tissue is not just an inert storage site; it's an organ that secretes a number of important hormones. Adiponectin is one of them, increased during prolonged exposure to cooler temperatures.[50] It promotes insulin sensitivity, glucose uptake into muscle (thus helping lower blood sugar), and the burning of fat. It also reduces inflammation, and as a result may help ward off inflammatory conditions like heart disease, cancer, and Alzheimer's disease.[51] In regards to the latter, adiponectin may also help improve insulin signaling in the brain.[52] This is notable, as Alzheimer's disease, sometimes called "type 3 diabetes," coincides with impaired brain insulin signaling.

Brown fat is good fat, and the more you have on your body, the greater the benefit you reap by activating it. Thankfully, gaining more is easy: embrace cooler temperatures. To prove this, researchers convinced healthy subjects to sleep in a temperature-controlled testing center for four months. For the first month, all rooms were set to 75°F, a temperature for which the body does not have to work to produce heat. For the second month, the temperature was lowered to 66°F. For the third, back to baseline, and for the fourth, up to a warm 81°F. Metabolic testing revealed that levels of brown fat increased during the 66°F month by an astonishing 30 to 40 percent.

One caveat: the brown fat dissipated during the warmer months, suggesting that routine cool temperature exposure is more beneficial than a "one and done" approach.

It's Getting Hot, Hot, Hot

Like cold, exposure to extreme heat was commonplace during our considerable time as hunter-gatherers. Consider persistence hunting, one of the oldest strategies employed by humans to catch elusive prey. Such endeavors would require a combination of physical endurance and prolonged exposure to the elements. Just as we have to warm ourselves, we've evolved numerous mechanisms to keep us cool and prevent the harm that could come from overheating. You're likely most familiar with sweating, which helps to maintain healthy body temperature as the perspiration evaporates from our skin.

Today our thermoregulatory systems are kept largely dormant, save for leisure-time activities like exercise, hot yoga, or sauna bathing. But allowing these systems to gather dust may undermine an opportunity to grow more robust and resilient, honed by millennia of our ancestors' struggles. Sauna therapy in particular—where one sits in a room heated by coals or electric coils—has been very useful for isolating, and studying, the health effects of heat. Much of the research on them comes out of Finland, the sauna capital of the world, where "taking a sauna" is as common as taking a shower.

Jari Laukkanen is a cardiologist from the University of Eastern Finland and a noted expert on the health effects of sauna. Using data from an ongoing study of heart disease in the Finnish population, he discovered that time in the sauna is associated with dramatically better health, including reduced risk for cardiovascular disease, dementia, and early mortality. And the more one partakes, the higher the risk reduction (this is known as a dose-response and is usually a sign of a causal link). In the case of Alzheimer's disease, for example, using sauna four to seven times per week was associated with a remarkable 65 percent risk reduction over twenty years.[53]

Below is a roundup of some of Dr. Laukkanen's other remarkable findings.

SAUNA RISK REDUCTIONS (Laukkanen et al.)

Times per week	0–1	2–3	4–7
High blood pressure	0%	24%	46%
Stroke	0%	14%	61%
Dementia	0%	22%	66%
Alzheimer's disease	0%	20%	65%
Early mortality	0%	24%	40%

Astonishing as these figures are, such findings do not *prove* that sauna is directly responsible—they could simply reflect the health benefits of more relaxation time. Nonetheless, the work of Dr. Laukkanen and others suggest that, as with cold, extreme heat does throttle the body's stress response in a beneficial way.

The next time you find yourself in a sauna or steam room, try placing two fingers over the radial artery in your wrist. You may notice an elevation in heart rate similar to a brisk walk or jog on the treadmill. This is because heat stress can be similarly taxing on the body as exercise, acting in many ways as an aerobic exercise *mimetic*. Heat also increases nitric oxide, a gas that dilates your blood vessels and increases blood flow throughout your body (you may feel this locally when applying a warm compress to a sore joint). The net effect is a lowering of heart rate and blood pressure and an improvement in arterial elasticity—all of which increase fitness and decrease the mechanisms underlying aging and decay.

FAQ

Q: Do hot tubs/steam rooms/infrared saunas work similarly to saunas?
A: It stands to reason that hot tubs, steam rooms, and infrared saunas would impart similar (though perhaps not identical) benefits as the sauna. One study, performed on young and healthy people, found that eight weeks of regular hot tub use also improved vascular stiffness, decreased arterial thickness, and reduced blood pressure similar to what we'd expect from sauna.[54] Still, the strongest prospective research has been performed in Finland with dry, hot sauna, making it the most trustworthy bet for reaping the benefits of regular heat stress.

These effects on your vascular system might be enough to explain why regular users are less likely to develop Alzheimer's disease, an incurable condition often preceded by dysfunction of the blood vessels that supply the brain. But saunas seem to fortify your health in other important ways. Similar to exercise, they cause a momentary increase in blood markers of inflammation. Don't be alarmed, though; this temporary "spike" in inflammation is key to acclimation, summoning the body's equivalent of a counterterrorism task force. It's why for regular sauna users, levels of inflammation tend to be lower.[55] In the case of sauna, the inflammatory challenge seems beneficial. It may also help keep your brain clean.

Proteins are important; they make up your cells, tissues, organs, and the countless worker chemicals that keep your body up and running smoothly. To function, proteins must fold into intricate, origami-like structures, but the stresses of modern life (including chronic inflammation) can cause them to misfold and become sticky. This is a cautionary tale for the brain, which has a tendency to accumulate protein-based plaques over time. Alzheimer's disease, for example, is defined in part by a buildup of plaque made up of the protein amyloid beta, along with "tangles" of another protein called tau. Tau tangles are so named for their tangled, misfolded appearance.

Thankfully, our proteins needn't contend with stress unprotected. Any significant stress activates guardian molecules called heat shock proteins. Though they are proteins themselves, they act like buttresses on a castle wall to protect other proteins from misfolding and may help prevent the brain from becoming a dumping ground for plaque. Sauna enlists a robust activation of heat shock proteins to the degree that in one study, two fifteen-minute intervals in a sauna heated to 208°F caused a roughly threefold increase in heat shock protein-encoding genes.[56] And repeated use (or becoming more fit in general) may lessen the pro-inflammatory "hit" induced by heat, while still providing the benefits of heat shock protein activation.

When sauna or other heat-based activities are made a regular part of your lifestyle, these factors may converge for a longer and healthier life, free of diseases like coronary artery disease and Alzheimer's. This may be especially good news for the healthy disabled, or anyone for whom vigorous exercise is difficult.

Feeling Warm and Fuzzy

Like exercise, saunas also boost your mood, and you may notice it after just one session. Many of the effects of the body's stress response are conserved across stressors—meaning whether it's cold stress, heat stress, or moderate exercise, the effects are often similar. The surge in norepinephrine (the chemical messenger involved in focus, attention, and mood) that you might experience during a dunk in an ice bath also occurs after a session in the sauna. But norepinephrine aside, saunas may possess an even more powerful mood-boosting mechanism up their sweaty sleeve: an influence on your brain's opioid system.

You've likely heard of feel-good chemicals called endorphins and their association with prolonged exercise. They're what underlie the "runner's high" that marathoners and other endurance athletes are fond of, due to their affinity for our brain's opioid receptors. Numerous prescription painkillers and illicit street drugs directly target these receptors, and these substances comprise some of the most addictive and dangerous compounds on Earth. Fortunately, regular sauna use causes a massive release of these proteins, imparting many of the euphoric effects without the considerable downsides.

Might saunas then be used to treat depression? In a pioneering study, Dr. Charles Raison of University of Wisconsin–Madison studied the effect of whole-body hyperthermia on the symptoms of major depression. When compared to a placebo treatment, a single thirty-minute sauna session significantly improved symptoms in people with major depression. The effect wasn't small either; Dr. Raison noted that the antidepressant effect was almost 2.5 times stronger than standard pharmaceutical treatment. What's more, benefits were sustained for six weeks afterward.

In tandem with endorphin release, saunas also call upon the ugly stepchild of the endorphin family, dynorphin. Unlike endorphin, which is primarily a euphoriant, dynorphin can induce a dysphoric effect, evoking feelings that sometimes accompany drug withdrawal. It's why prolonged exercise or time in the sauna can make you feel nauseous. Unpleasant as this sounds, dynorphin ultimately acts to our benefit; a temporary increase in dynorphin helps to increase receptors for endorphins, making you more sensitive to the feel-good effects of exercise or sauna.[57] The implication here is that to reap the full benefit from sauna (or exercise), you may be required to reach a point of discomfort first. The antidepressant effect may be stronger, and grow more pronounced with time.

Clean Air, Better Care

The final piece of the nature puzzle might seem obvious: breathing in clean air. Unfortunately, dirty air puts you at risk for premature death, reduced cognitive performance, and other serious effects. This has been a major concern for me, as I spend a lot of time in big cities breathing in air of dubious quality, and I'm not alone. More than 166 million people in the United States—52 percent of all Americans—are exposed to unhealthy levels of outdoor air pollution. You likely are one of them.

There are two main types of pollution in the United States. One is ozone, the other is particulate matter. The latter type of pollution refers to particles that are airborne, which gives them the ability to enter our lungs and circulate. The most dangerous types of airborne particles are those measuring 2.5 micrometers or smaller (PM2.5 for short). These particles, measuring about 3 percent of the diameter of a human hair, are invisible to the naked eye. They come from power plants, industrial processes, vehicle tailpipes, wood stoves, and wildfires.

One tiny PM2.5 particle is called magnetite. Magnetite is made of iron and is commonly found in the air of big cities. When we breathe it in through our noses, it's able to enter the body and travel up to the brain. Once there, it can "infect" multiple regions, including the memory-processing hippocampus, which is among the first structures to be damaged in Alzheimer's disease. Shockingly, these nanoparticles have been discovered in the brains of children as young as three, causing inflammation and impaired cognitive function.

Once inside of you, these tiny foreign particles (magnetite among them) contribute to blood vessel dysfunction. In one trial, healthy people exposed to very high concentrations of PM2.5 had a steep decline in heart rate variability (an important measure of heart health) and an increase in heart rate. This type of vascular dysfunction contributes to heart disease, but it also plays a major role early on in the development of Alzheimer's disease and other forms of dementia, partly due to its effect on the blood-brain barrier.

The blood-brain barrier is a selective membrane that keeps your brain safe from chemicals found in your blood that might harm it, while allowing the transport of vital nutrients and fuels like glucose. It's a vascular network that you don't want to mess with, as disruption of the blood-brain barrier is linked to Alzheimer's disease, autism, multiple sclerosis, epilepsy, and Parkinson's disease. Unfortunately, PM2.5 has been shown to cause blood-brain barrier dysfunction in young people and even promote the appearance of the two hallmarks of Alzheimer's disease—amyloid plaque and tangled tau proteins—way earlier than Alzheimer's typically rears its ugly head.[58]

Might air pollution actually make you less intelligent? That's a question that has been plaguing Chinese researchers as the nation's rapid industrialization and lagging environmental policies have caused air pollution to become a major public health threat. In one study that involved more than twenty-five thousand people living throughout China, exposure to increased levels of air pollution were tied to lower verbal (and math, to a lesser degree) test scores.[59] The study also looked at the rate of change over time, and lengthier exposure correlated with larger drops in performance. One of the authors, a professor of health policy and economics at the Yale School of Public Health, told NPR that stricter air-quality regulation would lead to cognitive gains on par with an extra year of education for the entire population.

Many of China's cities grapple with pollution far worse than what is found in the United States, but we are not immune here in the West—far from it. One study that occurred across forty-eight states found high exposure to air pollutants increased the risk of cognitive decline in women by 81 percent and Alzheimer's by 92 percent, compared to those living in areas with low exposure.[60] What's scarier, the researchers suggested a particular vulnerability in carriers of the Alzheimer's risk gene, ApoE4. They posited that one fifth of Alzheimer's cases may be owed to air pollution alone.

Not living in a polluted area may be the most obvious way to avoid exposure to potentially harmful pollutants, but let's be honest, that's not always practical. Plus, there are many things to be gained living in the world's great cities. Here, then, are some additional measures we can take to protect ourselves:

▶ **Eat your B vitamins, or consider a supplement.** In a small trial, healthy people exposed to high levels of PM2.5 for two hours experienced vascular dysfunction and inflammation.[61] They then took a daily B complex for four weeks (containing 2.5 mg of folate, 50 mg of vitamin B_6, and 1 mg of vitamin B_{12}), which seemed to completely protect them when the exposure was repeated.

▶ **Eat omega-3-rich foods, or consider a fish oil supplement.** Omega-3s both prevented and treated the inflammation and oxidative stress caused by air pollution in mice exposed to PM2.5, adding up to a 30 to 50 percent reduction in harm.[62] A trial involving elderly individuals yielded similar results.[63]

▶ **Eat your crucifers, especially broccoli sprouts.** Raw cruciferous vegetables, and especially young broccoli sprouts, produce sulforaphane, a potent activator of the liver's phase 2 detox enzymes that help you to excrete environmental toxins. In one study, sulforaphane (a cup and a half of raw broccoli daily for four days prior to exposure) negated the inflammatory effects of diesel exhaust by stimulating these antioxidant enzymes.[64] In another, sulforaphane significantly increased the metabolism and excretion of the carcinogenic gases benzene and acrolein.[65]

▶ **Load up on antioxidant-rich foods.** Six months of supplementation with vitamin E and C (800 mg and 500 mg, respectively) was effective in decreasing markers of lipid and protein damage and improved antioxidant defenses for people regularly exposed to coal burning.[66] Foods high in vitamin C include kale, berries, broccoli, and citrus fruits. Foods high in vitamin E include almonds, avocado, and extra-virgin olive oil.

▶ **Know your ApoE4 status.** Any consumer gene-testing site can find this for you (look to http://maxl.ug/TGLresources for specific recommendations). Carriers of either one or two copies may choose to spend less time in polluted areas.

It's clear that outdoor air pollution can be dangerous, but when was the last time you considered the air quality in your own home? You may be shocked to learn that indoor air can be as much as ten times more polluted than the outdoor environment. The surprising reasons for this—and what you can do to protect yourself—are going to be covered in chapter 5.

We are a part of a vast, interconnected, natural web, and have been for some time. Though our insulation from the forces of nature can be a welcome evolution to a sometimes cold and violent world, it has come at a cost. Exposure to the sun, natural variations in temperature, and fresh, clean air are key to a long and healthy, disease-free life. Whether you live in a big city or a small town, prioritize nature and integrate its teachings into your life.

Up next, the virtues of exercise.

FIELD NOTES

▶ Healthy vitamin D levels are not optional; get levels tested by your doctor to ensure that they hover between 40 and 60 ng/mL.

▶ If you are an older person, have a darker complexion, are over-weight, or live at a higher latitude, you may need either more time in the sun or a higher dose of vitamin D to attain a healthy level.

▶ If you choose to supplement, go with vitamin D_3 (as opposed to D_2), which is identical to what we create in our skin.

▶ The potential for vitamin D to improve brain and cardiovascular health, reduce risk for cancer, and improve or prevent autoimmunity is a growing interest in the scientific community with enough research to merit optimism.

▶ Cold stress (either via open water swimming, ice baths, cold showers, or cryotherapy) can be a potent means of enhancing vigilance and mood.

▶ Cold stimulates nonshivering thermogenesis, which can enhance metabolic health and burn calories—no treadmills required!

▶ Heat stress boosts heat shock proteins, which act to buttress other bodily proteins, possibly playing a role in the prevention of certain neurodegenerative illnesses.

4

GET OFF YO' A**

Be strong to be useful.

—GEORGES HÉBERT

Life is like riding a bicycle. To keep your balance, you must keep moving.

—ALBERT EINSTEIN

Is there any area of life where stagnation is desired? Certainly not in our professional lives! Who wants to work for the same boss, hold the same role, or make the same amount of money forever? Hopefully not in our personal lives, either. Wouldn't you love to make new friends or deepen your connection to your significant other? Stagnation is the opposite of a Genius Life; it is death and decay.

The human body doesn't like stagnation, either; it is designed to move. Yet civilized life in the twenty-first century brings with it an epidemic of inactivity. More than one third of US adults do not engage in any leisure-time physical activity.[1] For even the most active among us, exercise needs to be scheduled in order to fit it into the overwhelming demands of modern life. In this chapter, as we delve into the countless benefits and nuances of movement and exercise, you will begin to understand that for our health and well-being, stagnation is a toxin no different from refined grains and industrial oils.

We will focus on five different types of activity: non-exercise physical activity, aerobic exercise, high-intensity interval training, resistance training, and recovery. If this seems like a lot, don't worry; some of these activity types you're likely doing already. So whether you're a "noob" in the gym, a certified homebody, or a seasoned lifter chasing gains, the following pages will help you improve what you see in the mirror while laying claim to better health and a happier and more capable brain.

Get Off Yo' A**

Perhaps the most underappreciated type of activity is the physical activity that you perform every single day as you move about the world. This type of easy, absentminded activity includes walking to lunch, pacing while on the phone, typing, cleaning the house, cooking, dancing, climbing stairs, and folding laundry. It's what you're doing when you're not deliberately exercising, sleeping, or watching *Curb Your Enthusiasm* reruns from the comfort of your couch. We call this non-exercise physical activity.

When compared to the kinds of exercise you do in a gym, non-exercise activities may not seem very notable, but don't write them off. Cumulatively, they provide big health wins, especially when compared to sitting all day. One major benefit is that despite minimal perceived effort, they burn anywhere between 300 and 1000 calories daily—an attribute called *non-exercise activity thermogenesis*, or NEAT. Here are some ways that NEAT can show up in your life:

Dancing	Folding laundry
Walking the dog	Using a standing or treadmill desk
Cleaning the house	
Playing a musical instrument	Carrying groceries
	Doing the dishes
Typing	Shoveling snow

Singing	Playing with your kids
Having sex	Chasing your cat

Added up, these activities can account for about half of an active individual's daily energy expenditure.[2] Performing yard work, for example, or cleaning and maintaining your house, each can expend ten to fifty times more energy than sitting in front of the television. NEAT is so effective at dissipating energy, in fact, that it's why we tend to increase spontaneous movements when overfed.[3] Extra fidgety? Perhaps you ate too much for lunch.

One group from the Mayo Clinic decided to find out whether NEAT alone could prevent a person from becoming overweight. They gave normal weight subjects an additional 1000 calories' worth of food—the equivalent of a Triple Whopper sandwich—every day for eight weeks and documented their activity. As expected, overfeeding increased their daily activity, but those with higher levels of NEAT (washing the car, fidgeting, or playing with kids, for instance) were also able to preserve their leanness and prevent weight gain to a remarkable degree: the increase in activity was able to explain a tenfold variation in fat gain between participants.[4]

Calorie burning aside, low-intensity activity is able to influence fat storage via its effect on an enzyme called lipoprotein lipase, or LPL.[5] LPL (not to be confused with LDL, a cholesterol carrier) is found within your blood vessels and helps determine where the fat you eat ends up. When you eat foods such as a grass-fed burger or a handful of nuts, the fats from these foods get distributed on little rafts called lipoproteins. LPL allows various tissues to pluck the fats from these rafts so that they can either be used for fuel, in the case of muscle, or stored as fat. Simple activities—walking around the office, looking after kids or pets, or preparing food, for example—all increase LPL in the muscles being engaged, making fat storage less likely.[6]

FAQ

Q: I work a desk job. What should I do to achieve more NEAT?

A: Try using a standing desk and alternate between sitting and standing. If you don't have a standing desk, you can make one. I wrote much of this book on my dining room table and would often stack empty boxes and coffee table books to elevate my laptop. You can then put a leg up on your chair to open up your hips, switching sides in five-minute intervals. Take regular breaks to stretch or do a lap around the office and opt for the stairs whenever you can, even if it's just a two-minute break every half-hour (an interval shown to normalize blood flow to the brain, which diminishes with prolonged sitting[7]). And, while long deliberate bouts of moderate-intensity aerobic exercise are unnecessary if your lifestyle includes adequate NEAT, adding some cardio to your workouts can plausibly compensate if you truly are sedentary all day. I will offer some examples momentarily.

After eating, LPL tends to be lower in muscle and higher in fat tissue.[8] This occurs most dramatically after high-carbohydrate meals and snacks that stimulate insulin, and is partly how that ancient hormone promotes fat storage. Today we consume roughly 300 grams of carbohydrates per day, eating and snacking while sitting around at our desks, on the couch, and in our cars. As researchers in the *Journal of Applied Physiology* observed, "most individuals spend the majority of their day in the postprandial (post-meal) state, when physical activity is low."[9] This keeps fat burning effectively turned off, allowing our waistlines to expand ever outward.[10]

People with higher levels of activity aren't just leaner. The lipoprotein rafts travel through your arteries carrying fats in a particular form called triglycerides. Elevated fasting triglycerides (i.e., more fat in the blood) are associated with metabolic syndrome. This is usually a sign that your body is becoming overloaded with food energy and that you aren't using it effectively. Today "good" metabolic health is achieved by only 12 percent of adults.[11] That's unfortunate, since poor metabolic health shortens your life span and puts you at risk for any number of conditions, including heart disease and cancer.[12] Simply moving more—adding a daily walk to your routine, for example—allows LPL to reduce circulating triglycerides, which can lower your risk for heart disease.[13]

The life-extending properties of low-intensity activity were shown in a study involving nearly four thousand Swedish adults who were followed over thirteen years. Those who were most active had fewer cardiovascular events (such as heart attack) by 27 percent and reduced risk of early mortality by 30 percent.[14] On the one hand, this might prove the obvious—that healthy people move more and sick people move less—but studies show that simple movements like walking and standing spread out over the course of the day do improve markers of heart health, *even* in comparison to deliberate exercise for otherwise sedentary people.[15]

NEAT FOR THE AGED, FRAIL, OR DISABLED

NEAT provides a meaningful way of maintaining energy expenditure for people for whom vigorous exercise is difficult, such as the elderly, weak, or disabled. As my mother's Parkinsonian symptoms advanced, bringing on stiffness and balance problems, it became increasingly difficult for her to perform high-intensity exercise, making NEAT all the more valuable. I would often join her on walks or dance with her in her living room to her favorite music, holding her hands so she wouldn't fall (some of my favorite memories). And when it came to chores around the house, I wouldn't discourage her from doing them; I'd simply offer to help. Participate in NEAT with your loved ones, or volunteer at a local care facility. They will appreciate the time spent, however mundane it may seem to you.

Where your metabolism is concerned, movement is never insignificant and it's never pointless; it's key to staying trim and healthy in part because it helps to dissipate excess food energy. LPL is part of the ensemble cast responsible. Researchers for the *Journal of Physiology* wrote: "The striking sensitivity of muscle LPL . . . may provide one piece of the puzzle for why inactivity is a risk factor for metabolic diseases and why even non-vigorous activity provides marked protection against disorders involving poor [energy] metabolism." And it may even make your brain work better, too.

Sitting for an extended period of time literally causes blood to drain from your brain, and this can impair cognitive function.[16] But even simple movements (a two-minute walk for every thirty minutes of seated time) can promote normal blood flow. It works like this: when you move your body, you create small changes to your blood pressure, which pushes blood and nutrients to your brain.[17] Over the long term, reduced blood flow may be to blame for the mental foibles we usually consider part of aging, and it's also a suspect in both Alzheimer's disease and vascular dementia. In the latter, blood flow to parts of the brain is impeded by a series of tiny blockages, but one small study found that walking a mere three times a week improved cognitive function in patients with an early and mild form of the condition.[18]

Blood isn't the only bodily fluid reliant on movement to keep flowing. Muscular contractions, however slight, help to lubricate your joints, drain extracellular water, and push lymph fluid through their channels, which don't have the ability to contract themselves. These ducts lie in between muscle and skin, where they serve as the service roads for nutrient delivery and your hard-working immune cells. We now know that they are also connected to the brain, and provide an outlet for cerebral waste products like amyloid beta, among others. And just like standing water, stagnation of this fluid does not lead to good things.

To see whether simple daily movements could stack up to better brain health, researchers measured activity levels in older adults and matched them to their annual performance on cognitive tests. Published in the journal *Neurology*, the investigators found that higher levels of daily movement were linked to better thinking and memory skills. The observation held true even when postmortem analysis revealed Alzheimer's-related pathology (plaque buildup, for example) in the brains of subjects.[19] Everyday activity, it appears, helped fight off dementia.

NEAT proves that gym memberships, while nice to have for certain types of exercise, are nonessential to gaining or maintaining health, and certainly to losing weight (though, of course, paying for a gym membership can be a good motivator). Getting fit, shedding weight, and shielding yourself against an occasional high-calorie indulgence might be as easy as simply moving more.

Earn Your Carbs

Take a deep breath. Feel that? You are functioning in a state of aerobic respiration. Walking around, running errands, climbing stairs, chasing your cat—these activities are all fueled by a mixture of fat, sugar, and oxygen, all of which usually are around in sufficient supply (the word *aerobic* means "with oxygen"). Aerobic *exercise* simply implies moving up the intensity ladder, taking your normal energy-generating processes and kicking them up a notch.

Similar to NEAT, most forms of aerobic exercise can be sustained for a long period of time. Usually performed at a mild to moderate intensity level, aerobic exercise includes jogging, riding your bicycle, skateboarding, and swimming. You'll notice when performing these activities that they increase your heart rate and demand deeper breathing. When our muscles' energy needs *exceed* the rate at which oxygen can be delivered—i.e., when we perform bursts requiring our maximum effort—our muscles then switch over to an alternate energy state called *anaerobic* metabolism. Anaerobic means "without oxygen," and it can *only* burn sugar. But where does that sugar come from?

When it comes to sugar storage, our bodies are like an NYC apartment; there just isn't a lot of space. Your liver can store some sugar (about 100 grams), and your muscles generally store the rest. In total, your muscles can hold about 400 grams of sugar, give or take depending on how much muscle you have.[20] This may seem like a lot, but that's only about 100 teaspoons of sugar, divided up between your back, chest, biceps, triceps, glutes, and every other muscle you have. This stored sugar is known as glycogen.

Under everyday conditions when energy demand is low, like when you're sitting at your desk, that sugar mostly stays put. Imagine your average person, sedentary, consuming about 300 grams of carbohydrates daily—an easy feat when your diet revolves around processed grain products *and* you drink your sugar in the form of soft drinks and juices, like most Westerners today. That person is always running on a full tank with every meal merely topping off the already-full reservoirs in their liver and muscles. But resistance training and high-intensity exercise both help to burn through that sugar, making room for starches and sugars that may find their way onto your plate.

This is why if you are regularly performing long duration, high-intensity exercise such as vigorous workout classes or training for a race or a competition, it may be helpful to consume carbohydrates along with protein around your workouts to keep your energy up. As you've learned in chapter 1, we are at our most insulin sensitive in the daytime, which means there is an easier entry of sugar into muscle. On the other hand, the post-workout window, day or night, provides the added benefit of *insulin-independent* glucose uptake, which means that your muscles will literally *soak up* sugar from your blood, reducing the requirement for insulin and turning you into a high-efficiency fat-burning beast.

FAQ

Q: How much protein should I consume?

A: Consuming adequate protein is an important part of growing and maintaining muscle, and this becomes even more critical with age. The most current evidence suggests that a minimum of .7 gram of protein per pound of lean body weight per day is optimal to grow and maintain lean mass while on a weight-training regimen.[21] For a 135-pound athletic female, this means 95 grams of protein per day. For a 185-pound lean male, this equates to 130 grams of protein per day. If you are carrying extra body fat, use your *goal* weight, and multiply by .7. High-protein foods to consider incorporating at each meal include grass-fed beef, eggs, chicken, turkey, pork, fish, and full-fat grass-fed Greek yogurt (fat-free yogurt is a good option, too—just avoid sugar-sweetened varieties). And if you work out while fasted, be sure to consume protein (such as a whey protein shake) soon after.[22] Feel free to revisit page 28 for a recap on other protein benefits.

If you're an advanced athlete, carbohydrates are important for strength and power, but if you're not, you may be able to skip them. The body is able to produce its own sugar, and glycogen will naturally refill over time, particularly as you begin to eat more plants. Ultimately, whether or not you need post-workout carbohydrates depends on your health and your goals, but as a general rule, higher-intensity workouts with greater regularity equals greater carb tolerance (see page 214 for a guide). Going carb-free after exercise can impart its own set of benefits, too, such as sustained fat burning and elevated growth hormone, which helps to strengthen joints and rebuild connective tissue.[23] Our bodies are highly adaptable and individual, so feel free to gauge your own carbohydrate requirements by alternating low- and high-carb meals after working out and seeing how you feel.

Boost Your Fitness

The more fit you are, the more effort you can give to your workouts while staying on the aerobic, fat- and oxygen-burning side of metabolism. We can measure where this threshold lies for each of us with a test of VO_2 max. This test describes the maximum amount of oxygen a person can utilize during intense exercise. Usually in a gym setting, a fitness professional will ask you to perform stationary exercise at increasing intensity while wearing a mask that analyzes your breath. Somebody with a higher VO_2 max will more efficiently handle oxygen, and therefore be able to perform at a higher intensity for longer. We call that endurance.

The threshold where you'd switch over to sugar-burning metabolism is called your lactate threshold. Ever hear of lactic acid? It's a well-known harbinger of muscle exhaustion and the source of the infamous "burn" during intense exercise. It can also lead to the temporary dysphoric feeling immediately following a hard workout. Previously considered a waste product of metabolism, we now know that lactate is actually produced all the time in our muscles and serves as a potent fuel source for the heart, brain, muscle, and other tissues. Being more fit means you have a greater ability to generate and utilize lactate, thus enabling higher performance.[24]

LACTATE FOR BRAIN HEALTH

Thanks to lactate research pioneer George Brooks, we know that lactate produced in muscles can enter circulation to fuel other organs, including the brain, where it surfs in on the same transporters that allow ketones entry.[25] At rest, the brain may derive about 10 percent of its energy from lactate, and vigorous exercise, the chief contributor to circulating lactate, pushes even more lactate into the brain.[26] Like ketones (described on page 15), lactate displaces sugar, the brain's usual fuel. This is good news because some brains have difficulty utilizing sugar to generate energy, including those with Alzheimer's disease, carriers of the common Alzheimer's risk gene (the ApoE4 allele, shared by 25 percent of people), and sufferers of traumatic brain injury.[27] Some research suggests that lactate may even be protective against Parkinson's disease; one trial showed that a high-intensity exercise routine prevented progression of the disease for patients after six months, while moderate intensity did not.[28] More reason to get off your a**!

Conventional wisdom tells us that to improve our fitness, we must perform long bouts of "cardio." While extended cardio sessions can boost endurance, high-intensity interval training, or HIIT, may be as effective at boosting your VO_2 max, in less time and with less collateral damage to your knees and joints. One study published in the journal *PLOS ONE* found that high-intensity interval training, performed three times per week over twelve weeks, imparted the same improvements in cardiorespiratory fitness and endurance as steady-state cardio in one fifth of the time.[29]

Training intensely—so intensely that you can only sustain it for brief bursts—is the equivalent of sending a text message to your genome that your cells either need to keep up or die. Your cells don't want to die, so instead they adapt so that they can churn out more energy. One such adaptation is mitochondrial biogenesis, or the creation of new mitochondria. These structures are housed in your cells and are where ATP, or energy, gets produced. More energy-producing mitochondria means more energy, and just one single HIIT session has been shown to be sufficient for stimulating this energy-enhancing process.[30]

FAQ

Q: When is the best time of day to work out?

A: Whichever time is most convenient for you! While there is a circadian (i.e., rhythmic) influence on our strength (people tend to be stronger later in the day), ultimately, the best time to work out is the time that you work out. Personally, I often feel strongest in the morning on an empty stomach, but I also enjoy late-afternoon workouts and following them with a large, protein-rich dinner. And while exercise in the evening (before 10 p.m.) has been shown to boost sleep quality, try to avoid working out *immediately* before bedtime, as doing so may have the opposite effect.[31]

To establish a HIIT routine, begin with three sets of twenty-second all-out reps of your exercise of choice, each separated by one or two minutes of low-intensity recovery movement. Some examples of this might be twenty seconds of vigorous cycling on a stationary bike with a period of slow cycling, and repeating. Or you may prefer getting out of the gym and running sprints up a hill for twenty-second intervals. In my gym, I often use the Assault AirBike or swing the heavy "battle ropes" around for three to five repetitions at the end of my workouts. The rules are loose, but the key to HIIT is to push your body just past its aerobic limits, recuperate, then repeat. Here are some examples of HIIT exercises:

Jump squats	Power yoga
Hill sprints	CrossFit
Burpees	Jujitsu
Bike sprints	Spin class
Battle ropes	Boxing

Remember: your "all-out" will be different from your sister's, your brother's, or your best friend's. Push yourself to your own personal limit, and then dial back. What matters most is the effort you give it, not outcomes like speed or distance. Listen to your body. And it's always a good idea to check with a doctor first, especially if you have a medical condition. As your fitness increases, feel free to add additional resistance (i.e., a steeper incline) or number or duration of repetitions.

Boost Your Brain

Where your brain is concerned, exercise is a must. Some of the benefits of a good workout are incurred directly, such as the fuels and nutrients and various neuroprotective chemicals that flood your brain with vigorous activity. Others are indirect but no less important, such as the protection exercise provides to your blood vessels.

The blood vessels supplying fuel, oxygen, and nutrients to your brain are under constant attack today. In chapter 1, you discovered how inflammation and chronically high blood sugar is akin to taking a blowtorch to them; probably not high on your list of things to do. But high blood pressure is another major threat, and a common one. About one in three adults suffer from hypertension (the medical term for high blood pressure), and this is not just a problem for your parents. According to the Centers for Disease Control, about 14 percent of twelve- to nineteen-year-olds now either have hypertension or are on their way there.[32]

Similar to inflammation or high blood sugar, you can't feel whether or not you have high blood pressure—it is called the silent killer, after all. But high blood pressure doesn't just increase your risk of an early death. It can damage your kidneys, your eyes, your sex organs, and your brain. Different forms of cognitive impairment can result from chronically high blood pressure, including dementia and a form of "pre-dementia" called *mild cognitive impairment*, or MCI. In one large multicenter trial, cognitively healthy people who had high blood pressure were less likely to develop MCI when they lowered their blood pressure with medication.[33]

But what causes high blood pressure to begin with? Stress is a big contributor, as are diets that include a lot of sugary beverages or processed foods.[34] This is partly because sugar excites the nervous system, eliciting a biochemical stress response even when you're perfectly calm. Studies show that even one sugary drink can raise your blood pressure for up to two hours after ingestion.[35] Another contributor to high blood pressure is obesity; if you're overweight, losing weight can often help correct blood pressure problems. And then there's physical inactivity.

You won't see TV ads asking you to talk to your doctor about it, but a robust head-to-head analysis published in the *British Journal of Sports Medicine* found exercise to lower blood pressure as effectively as medication for people with high blood pressure.[36] Exercise is also now an official treatment guideline for MCI under the American Academy of Neurology, which is welcome news, since there is currently no FDA-approved pharmaceutical treatment for the condition. The current recommendation for people with MCI is 150 minutes of moderate intensity aerobic exercise per week for six months, which may help slow or reverse cognitive decline.

If you are sedentary, beginning an exercise routine should bring about noticeable improvements in your cognition.[37] Even a single twenty- to thirty-minute exercise session may give you a learning advantage by activating regions of the brain involved in executive function and memory processing.[38] One study involving college students found that exercise before and during foreign language lessons promoted stronger recall and better integration of learned concepts.[39] The exercise needn't be backbreaking, either; moderate intensity aerobic exercise, at 60 to 70 percent of maximal heart rate, seems to do the trick.

NERD ALERT

Exercise enhances your all-day brain function, but it may be particularly useful while you are actively trying to learn. Since taking academic classes from the gym isn't really an option, try stacking your smartphone with talks, podcasts, or lectures as you engage in your preferred mode of exercise; you may be surprised by the results.

Aside from memory function, fitness has been associated with improvements in anxiety and depression symptoms. The World Health Organization estimates that as many as three hundred million people now suffer from depression, and the evidence connecting depression to low fitness has never been stronger. In a study of one million people, low cardiorespiratory fitness was associated with a 75 percent higher risk of depression.[40] Depressed people may choose to exercise less, leading to lower fitness, but a growing body of research suggests that when patients improve their fitness, their symptoms of depression improve, too.

One neurotransmitter thought to be involved in a balanced mood is serotonin. You may have heard of serotonin and its association with selective serotonin reuptake inhibitors (SSRIs), a common type of antidepressant drug. SSRIs work—for some—by increasing serotonin availability in the brain. SSRIs tend to be overprescribed, but it has been shown that their efficacy increases with the severity of depression. Though they can relieve some symptoms of depression, they are frequently associated with unpleasantries like loss of libido and suicidal thoughts, begging the question: what if you could increase serotonin on your own, without risking such side effects?

Serotonin is created in the brain from an amino acid called tryptophan, which we get from protein-containing foods. Tryptophan is carried into the brain by transporters along the blood-brain barrier. Under normal circumstances, other amino acids vie for the same transporters, but during exercise, tryptophan cuts to the front of the line, flooding the brain with serotonin. This is just one reason why exercise makes us feel great. And as a bonus, the increased availability of tryptophan to the brain may also help us sleep better. Later in the day, serotonin is converted to melatonin, the sleep hormone.

EXERCISE: A WIN FOR DEPRESSION AND ANXIETY

In one recent meta-analysis (study of studies) published in the journal *Depression and Anxiety*, eleven trials comprising 455 patients were analyzed. The study found that the antidepressant effects of aerobic exercise were significant across a large spectrum of patients and treatment conditions, even in a short time frame—less than a month.[41] The researchers highlighted the importance of this nuance, since pharmacologic treatment usually requires *at least* four weeks for any effect to emerge.

Lifting weights was also found to help improve mood in two large meta-analyses of randomized clinical trials, one published in the journal *JAMA Psychiatry*. Whether or not someone felt formally depressed at the start of the study, strength training was linked to improvements in depressive symptoms.[42] And the effect was strong at relatively small "doses," imparting similar relief whether subjects worked out twice or five times a week.

The same researchers performed another meta-analysis, this time looking for an effect of weight training on anxiety. In this study, published in *Sports Medicine*, the scientists found that for both healthy people and those with either physical or mental illness, resistance training significantly improved symptoms of anxiety.[43] With the addition of bonus effects like improved health, willing yourself to the gym should always be part of your treatment. And if you still need medication, that's fine, too, and talking to a therapist is never a bad idea, either.

Favorably altering the brain's biochemistry is one perk of exercise, but exercise has also been shown to promote healthy brain volume over time, preserving and even increasing its size when it would typically decline. One of the chemicals responsible is *brain-derived neurotrophic factor*, or BDNF, which becomes increased during vigorous exercise. BDNF has been referred to as the brain's Miracle-Gro protein by laypeople and scientists alike, and for good reason; it helps keep the brain youthful by encouraging the growth of new brain cells in a process called *neurogenesis*.

BDNF is powerful. So powerful that when sprinkled on neurons in a petri dish, it causes them to sprout dendrites—the spiny structures required for learning—like a Chia Pet. As you can imagine, BDNF has drawn interest for its implications for memory disorders like Alzheimer's disease. In Alzheimer's, BDNF is decreased by close to 50 percent, making the brain less amenable to change and reducing the characteristic known as *plasticity*. This makes raising BDNF with exercise a potential weapon in the fight to treat or prevent such conditions and even cognitive aging itself.

Plasticity isn't just important for being able to create new memories; it's also crucial for a healthy mood. People with clinical depression have brains that are more resistant to change, offering a biological explanation for why those with depression can often feel like they're stuck in a ditch. Studies have shown that many antidepressant drugs (including SSRIs) also boost BDNF, and one theory is that by increasing the brain's ability to change, they help it to "rewire" old patterns and facilitate healing (serotonin is also critical for neurogenesis to occur, which is also increased by some of these drugs[44]). The good news is that for many of us, exercise works at least as effectively as antidepressants.

How much aerobic exercise should you be doing? Unfortunately, there is no one-size-fits-all answer to this question. Your ideal prescription will depend on your goals: Are you training to increase endurance? Or purely for better health and fitness? If your answer is the latter, you may not need to do any deliberate "cardio" if your life naturally involves lots of activity in the form of NEAT (for example, walking or carrying things). In that case, focus your efforts on resistance training, which is discussed next.

On the other hand, if you are mostly sedentary, deliberate aerobic exercise might be a reasonable antidote. In that case, aim for 30 minutes per exercise session and 150 minutes of total aerobic work per week. Ultimately, your ideal protocol will depend on numerous factors, but the most important of them is your enjoyment. And keep in mind, you may also tweak other forms of exercise, such as weight training, by simply shortening rest times, to help you reap the benefits of aerobic exercise and meet your goal quota.

Here are some examples of aerobic exercise:

Vinyasa "flow" yoga	Jogging
Hiking	Rowing
Riding a bicycle	Skiing
Swimming	Circuit training

CARDIO FOR WEIGHT LOSS?

Aerobic exercise has a lot of benefits, but its usefulness for weight loss is frequently overstated. While "cardio" does accelerate the burning of calories, non-exercise activity thermogenesis (NEAT, described earlier) eats up far more calories than most people would ever care to burn on cardio equipment. Excessive cardio, especially without resistance training, may even lead to muscle loss, hurting your metabolism and promoting the unfortunate condition known as "skinny fat." This exact scenario was put on display in a Wake Forest University trial involving older, dieting adults: aerobic exercise caused them to lose twice as much muscle when compared to those who lifted weights instead.[45]

The takeaway? Aerobic exercise does a number of spectacular things for mind and body, but to look better naked and have a buffer against the occasional treat, *regular* movement, weight lifting, and improving the quality of your diet are likely a better strategy.

Get Swole to Not Get Old

Regardless of your age or gender, putting on muscle should be a primary focus of your exercise routine. Aside from helping you move freely about the world and perform everyday tasks, resistance training (the primary means we use to grow stronger) increases the strength of your bones, aids in weight loss, lowers inflammation, and supports your metabolic health, burning calories while providing a buffer for occasional carb binges.

Research suggests that having stronger muscles is directly linked to better health as one ages. A University of Sydney study of eighty thousand people found that those who strength-trained on a weekly basis had a 23 percent reduction in risk of premature death by any means.[46] And no gym membership was required to gain the benefits: people who did body-weight exercises, like push-ups, sit-ups, and pull-ups, yielded comparable results to gym-based activities!

The research on resistance training and the brain is in its nascent stages, but physical strength does seem to be associated with better cognitive function among older adults, and it may even help one particularly desperate group: patients with cognitive impairment. One study in patients with MCI found that a six-month weight-training regimen led to significantly higher scores on cognitive tests compared to baseline—*and* they retained their improvement at twelve months. Those who had the greatest gains in strength had the greatest gains in test scores, while the control group who performed only stretching exercises saw a decline in their cognitive abilities.[47]

Unfortunately, age is still the top risk factor for developing Alzheimer's disease and other neurodegenerative conditions. Is it any coincidence then that age-related muscle loss is also common? You can lose as much as 3 to 5 percent of your muscle mass each decade after age thirty. The good news is that any time is the right time to begin a weight training regimen. Even elderly sedentary subjects can achieve more than a 50 percent strength gain after just six weeks of weight training two or three times per week.[48] The key at any age is to lift enough weight to challenge yourself (about 70 to 80 percent of your maximum strength).

When lifting, focus on *compound* movements. These types of

movements recruit muscle fibers from various muscle groups at once. For example, a bench press involves the muscles of the chest, triceps, and shoulders. A biceps curl, in contrast, is an *isolation* movement and primarily involves one muscle: your biceps. Isolation moves can be used to augment your workouts, but your focus should be on exercises that are going to give you the most bang for your buck, particularly if you're just starting out. Other compound exercises include squats, pull-ups or lat pull-downs, deadlifts, and shoulder presses.

Thankfully, there are many different exercises to suit your tastes. Weight lifting, calisthenics, and products like resistance or power bands are all great modalities that allow for compound exercises. These exercises are meant to be performed at or near maximal effort and in short bursts (typically called "sets"). Don't worry about going to failure, which means an inability to proceed to another repetition ("rep"), especially if you're new to lifting.

HAVE AN EXERCISE "SNACK"

When it comes to exercise, "snacking" is most certainly good for you. In a study examining the effects of different types of exercise on blood sugar, a mere six minutes of either intense uphill walking or walking combined with resistance training shortly before breakfast, lunch, and dinner led to lower twenty-four-hour average blood sugar concentrations compared to thirty minutes of continuous exercise before dinner.[49] The take-home message is this: "snack-sized" bursts of intense, sugar-burning (i.e., anaerobic) exercise may help manage your blood sugar better than longer bouts of moderate intensity aerobic exercise. No time? No excuse! Even a quick few sets of push-ups or air squats before a big meal can help your body better manage the ensuing flood of energy.

To build out a weight-training routine, plan on doing a full-body workout three times per week. You may decide that you enjoy doing a "split," whereby you exercise different body parts on different days, but some research suggests that for both beginners and advanced trainers alike, working the same muscle more than once per week is required to achieve progress in terms of strength and muscle gain.[50] After each strength-training session, you may include some high-intensity interval training, cardio, or even a session in the sauna, which has been shown to enhance exercise recovery. Here are some options for your routine:

Full-Body Routine

Monday

Squats	Shoulder presses
Stiff-legged deadlifts	Hammer curls
Lat pull-downs	Triceps push-downs
Bench presses	

Wednesday

Lunges	Barbell curls
Hamstring curls	Delt raises
Incline presses	Skull crushers
Pull-ups	

Friday

Repeat Monday

Split

Monday (chest/biceps)

Bench presses	Barbell curls
Incline presses	Hammer curls
Flyes	

Tuesday (back/triceps)

Pull-ups	Skull crushers
Wide-grip lat pull-downs	Triceps push-downs
Cable-assisted chin-ups	

Wednesday (legs/shoulders)

Squats	Shoulder presses
Lunges	Upright rows
Hamstring curls	Face pulls

How many sets and reps should you do? On a full-body day, you may pick one exercise per muscle group, whereas on a split, you may choose three or four. For each exercise, perform 3 to 4 sets. Regarding reps, there are benefits to be gained by incorporating a range. You will reap muscle growth across a range of repetitions, from 1 to 20, and for building strength, lower rep ranges (between 1 and 5) are the most effective. If you're new to training, begin in the 8 to 12 rep range. Lower reps will necessitate heavier weight and should only be approached once you have developed good muscular control and stability.

No matter where you begin, proceed slowly, focusing on good form. For true newbies, or if you're nursing an old injury, I highly recommend investing in a few sessions with a qualified personal trainer to show you proper form and technique. Some movements (particularly deadlifts and squats) can hurt you if you aren't careful. You only get one body; treat it kindly!

Rest Yourself

Rest is medicine for the active body and mind, but are you really resting when you're lying in bed, refreshing your Instagram feed? To truly rest today requires diligence to ensure that we are reaping its maximum benefit. When it comes to supporting your exercise routine, there are two major types of rest I will cover: sleep and relaxation. Each one has unique benefits not provided by the other.

The benefits of good sleep cannot be overstated. It helps lock in the gains made in the gym by allowing your nervous system to rest—critical for a routine that includes strength training and high-intensity interval training. And sleep acts as a master controller to your endocrine system—the hormones that control everything from tissue repair and development to energy expenditure to hunger.

One of the hormones that is increased during sleep is growth hormone (GH). Secreted by the pituitary gland in your brain, GH strengthens your connective tissue and helps maintain lean mass. Both are vital to prevent injury and to facilitate exercise adaptation so that you can "hit it" again tomorrow. Beyond exercise, GH also provides benefits for brain function. In one study, GH replacement therapy boosted cognitive function in patients with mild cognitive impairment and in healthy controls after just five months.[51]

Most of our growth hormone is released during slow wave sleep with the largest pulse occurring soon after sleep onset. With less than seven hours of sleep per night, growth hormone release is attenuated, causing our brains to release the hormone during the day as a compensatory measure.[52] But as I've mentioned, growth hormone is inhibited when we consume carbohydrate-rich foods—the exact types of foods we crave most when we're underslept.[53] This is also why late-night eating may infringe on growth hormone release, and is yet another reason why you should consider curtailing food consumption two to three hours before bed.

GROWTH HORMONE: A FASTING HORMONE

In adults, growth hormone serves to preserve lean mass during periods of famine or in a fasted state.[54] One of the best ways to boost growth hormone, therefore, is with fasting. When one fasts for anywhere from fourteen to sixteen hours onward for a female and sixteen to eighteen hours onward for a male, growth hormone begins to increase. After twenty-four hours of fasting, growth hormone has been reported to shoot up as high as 2000 percent! Just remember that everyone is different, and some women may have an *adverse* reaction from extended fasting (revisit page 66 for other caveats).

Soon after the release of growth hormone, another hormone called testosterone takes the stage. Testosterone may be most famous for male sexual development and behavior, but it's also important for the development and maintenance of muscle mass, strength, bone density, and well-being for both men and women. The majority of our daily testosterone release occurs during sleep, and sleep deprivation greatly hinders this nightly surge. In one study, young and healthy men who underwent one week of sleep restriction to five hours per night experienced a decline in testosterone by 10 to 15 percent.[55] Testosterone normally decreases by 1 to 2 percent each year, so to put that into perspective, in just one week of poor sleep, levels of one of the most powerful hormones in your body become equivalent to that of someone five to ten years older! Sadly, at least 15 percent of the US working population regularly experiences that degree of sleep restriction, willingly or not.

Both growth hormone and testosterone have been referred to as antiaging hormones (the latter more specifically for men), and this is partly because levels peak during youth and decline with age. Seniors in particular spend less time in deep sleep, and optimizing sleep increases natural growth hormone and testosterone production as a result. As you learned in chapter 2, age-related sleep decline may be partly due to circadian dysregulation caused by reduced sensitivity to light. No matter your age or stage, remember that good sleep habits begin with morning sun exposure, even on a cloudy day. (I'll give you more tips for optimizing your sleep in chapter 6.)

Aside from your sacred slumber, relaxation is an important and underappreciated component of fitness. Taking time off from your workouts is as essential as the workouts themselves! Excessive training with inadequate rest can and will result in overtraining, handicapping your performance, causing you to feel fatigued, and leaving you prone to injury and illness (revisit the concept of allostatic overload on page 82).[56]

Rest allows your muscles to grow, your joints and ligaments to strengthen, and your nervous system to adapt to the increased workload. It may help to know that strength is as much in the mind as it is in your glutes. In other words, your nervous system plays as much of a role in your strength as your muscles do.[57] This is key to the distinction between the way powerlifters and bodybuilders work out. While both are strong, bodybuilders have maximized their training for muscle size and have traditionally relied on lighter weights and higher repetitions. Powerlifters, on the other hand, train for strength using heavier weights to challenge the mind-muscle connection.

The take-home message here is that for increasing intensity, rest becomes crucial to the adaptation process. Studies show that muscle protein synthesis—or the growth of new muscle tissue—occurs for about forty-eight hours after exercise, meaning you should probably give yourself a good two days before attacking the same muscle twice. In the end, the specific rest needs of every person will vary. Some people can train with much higher volume and require less rest than others. This depends on myriad variables: age, sex, fitness levels, nutritional status, sleep, and others. As always, experimentation is key. Now go have fun!

SPEND MORE TIME ON THE FLOOR

The Western culture of chair sitting has caused an epidemic of weakness and tightness in our bodies, evidenced by widespread posture issues, back pain, and even pelvic floor dysfunction, which influences urinary and fecal control and sexual performance. Chronically being chair-bound causes your glutes (which should be the strongest muscles in your body) to get weak and your hip flexors to tighten. This can cause pain in the lower back, or leave it vulnerable, so that compensation in the gym may lead to serious injury. Spending time on the ground is common for children as well as adults in Asian countries and hunter-gatherer populations. It may strengthen many aspects of your body, including heart, spine, and digestive and reproductive health. Make it easy to sit and stretch on the floor by leaving a yoga mat out, or by buying a comfortable carpet for in front of your TV. To work from the floor, place your laptop on a chair and sit cross-legged (if difficult at first, you may prop your butt up with a towel, yoga block, or pillow as your muscles begin to strengthen).

Even basic movement helps to mobilize and flush waste-circulating fluids, lubricate your joints, strengthen your bones, oxygenate your organs, and push fresh blood and nutrients up to your ravenous brain. It also cuts risk of dementia, numerous types of cancer, and heart disease, the latter being the leading cause of death worldwide. Take these learnings with you for a leaner, healthier body, a longer life, and a brain that works and feels as well as it ought to.

Up next, the toxic everyday chemicals that—despite all your best efforts—may be holding you back from your true inner genius, and what to do to protect yourself.

FIELD NOTES

▶ Being idle causes blood and other fluids to pool and fat to build in your arteries. Instead integrate as much movement into your day as you can. You needn't kill yourself—this is low-intensity activity we're talking about, which can even be performed while sitting.

▶ Non-exercise activity thermogenesis (NEAT) includes walking, dancing, cleaning, and playing with your kids or pets. It is vital for health and cumulatively burns off a lot of excess energy—way more than a workout.

▶ High-intensity interval training provides a powerful stimulus to your muscles that says "Adapt!" This provides cellular antiaging benefits. The exercise must involve repetitions that are sufficiently intense to count. You can tell each repetition is sufficiently intense when you can only sustain the activity for twenty to thirty seconds.

▶ Aerobic exercise is important for brain health, mood, cardiovascular health, and endurance. While its utility for weight loss is overstated, it's still good for you. Just keep in mind that you don't need to be on the treadmill for hours, as it can be integrated into forms of exercise other than traditional cardio.

▶ Weight lifting is critical. Having greater strength is directly proportional to a healthier brain. It also allows your body to continue to exercise well into old age. This will help prevent frailty and the muscle wasting condition known as sarcopenia.

▶ As valuable as exercise is, rest is just as important. Sleep well and relax regularly.

5

TOXIC WORLD

You're eating healthy and making a point to sun yourself daily. You mind your body's circadian rhythm and commune with nature semi-regularly. You're active and exercise, and even suffer through the occasional cold shower. You've *finally* outsmarted the modern world, you think, extricating yourself from the conveyor belt that pulls the rest of us toward entropy and decrepitude. And yet, something still feels wrong.

When it comes to feeling good, being healthy, and performing your best, exercising, eating right, and getting outside is half the battle. Undermining all of our best efforts may be the chemicals that overlay, and indeed make possible, our comfortable, twenty-first-century lives. These include compounds we ingest, like certain over-the-counter medicines that when used routinely can have profound and unintended consequences to our health and well-being. But they also include the hidden chemicals that permeate our homes, finding their way into our bodies either through our skin, our food, or the air we breathe.

I became curious about these substances as a consequence of my mom's declining health; as her dementia advanced, she became increasingly homebound. I wondered how, for example, on top of the dozen ineffectual drugs she was taking, the caustic cleaning products her health aides used could be affecting her health. Due to their ubiquity in our personal environments, many chemicals are given carte blanche access to our bodies, where such compounds can linger for years—decades, even—at a time. And with North Americans now spending 93 percent of their time indoors, indoor air quality should be a major concern for us all.[1]

The following pages will unearth the potential toxicity wrought by the most common of these industrial compounds, including many cleaning products, foods, and even seemingly benign cosmetics and over-the-counter medications. For each, I'll include the actionable steps you can take to protect yourself and help your body to more effectively detoxify.

Endocrine Disruption

The endocrine system is the ultimate influencer; it orchestrates nearly every aspect of your life. It is made of long-range chemical messengers called hormones, which have receptors in cells throughout your body. In chapter 1 you read about insulin, one of the hormones in this system, which influences fat storage. But there are dozens of hormones that affect everything from sexual arousal to your predisposition for disease and even for various mental states. Here are some of the hormones you may already be familiar with:

Hormone	Partial List of Functions
Insulin	Fat storage and metabolism
Ghrelin	Hunger
Leptin	Energy expenditure, satiety
Cortisol	Stress, energy production
Testosterone	Muscular development, sex drive, sex organ development
Estrogen	Fertility, sex drive, sex organ development
Thyroid	Brain development, energy, metabolism

Hormones are powerful substances, and even a small change to the finely tuned endocrine system can have big effects. Put into perspective, if "normal" levels of a given hormone were able to fill up twenty Olympic-sized swimming pools, the addition or removal of just one drop of water could cause a response in the body. This gives hormones considerable power over you, and it also leaves you vulnerable to the consequences of their unintended alteration.

You can't feel or see it happening, but once inside your body, certain household and industrial chemicals can mimic various hormones by activating the cellular receptors for them. They can also prevent your own, naturally produced hormones from binding to their intended sites. Such chemicals are known as *endocrine disruptors*. According to the Endocrine Disruption Exchange, a science-based nonprofit dedicated to tracking and spreading awareness of these chemicals, there are over 1,400 substances with the potential to do this, and we encounter many of them every single day.

Part of what makes these chemicals so treacherous is that their activity is hard to predict. Nearly every chemical known to humans—even water!—becomes toxic at a high enough dose. But endocrine disruptors like the ones you will read about defy this logic, precisely because they act like hormones once inside the body. This allows them to affect the way your cells function at a dose far lower than that at which they become overtly toxic.

Most of the chemicals that we worry about follow the usual "dose makes the poison" paradigm—higher dose, higher chance of toxicity. For many decades, toxicologists did not think low dose toxicity was possible, and the idea that tiny amounts of an endocrine disruptor can still be dangerous is controversial. Add to the confusion the hundreds of billions of dollars spent to produce and market the products made with these chemicals, and you can see how they've been able to dodge scrutiny.

As I've mentioned, hormones govern a lot in our bodies, and so chemicals that perturb them can influence our susceptibility to a whole host of problems, including weight gain, metabolic diseases, infertility, and certain cancers. When we're young and still developing, exposure can have lifelong implications. The list of adverse health outcomes associated with endocrine disruptors therefore also includes genital abnormalities, endometriosis, early puberty, asthma, immune disorders, and even attention-deficit hyperactivity disorder (ADHD).

While it would take volumes to detail the hiding places of all potential endocrine disruptors, the most common are found in plastic.

Plastic Paradise?

The two most common and well-studied plastic compounds are phthalates (pronounced "THAL-ates"—the *ph* is silent) and bisphenols. Generally speaking, bisphenols are used to make plastics hard, and phthalates are used to make plastics soft. Bisphenols can be found in furniture, baby bottles, the lining of cans, plastic cutlery, and writing utensils. Phthalates are usually found in single-use plastic bottles, takeout containers, plastic storage containers, apparel, industrial tubing, and straws. These chemicals aren't always in plastic, either. Store receipts—the kind you can write on with your fingernail—are covered with bisphenol, and synthetic fragrances used in cleaning and personal care products are often created using phthalates.

Of the bisphenols, bisphenol A, or BPA, is the most well known and is commonly associated with food packages and reusable water bottles. Increasing consumer concerns around BPA has led many manufacturers to remove it from their products and label them "BPA-free," but that doesn't mean these products are free from related compounds. Some manufacturers are now using bisphenol S (BPS). Far less research has been done on BPS, but it likely has similar effects on the body to BPA. "It's created a dangerous game of chemical Whac-A-Mole," Dr. Carol Kwiatkowski, the executive director of the Endocrine Disruption Exchange, told me.

It began in the early 1900s when researchers were looking for a hormone replacement that would alleviate menstrual cramps and symptoms of menopause and pregnancy (hot flashes and nausea, for example), and could be used in the prevention of miscarriages. In the mid 1930s, a medical researcher at the University of London named Edward Charles Dodds discovered a candidate in a chemical that had been synthesized in Germany thirty years prior. It was bisphenol A, and it seemed to mimic the female sex hormone estrogen.

An estrogen replacement would add tremendous value to society; it would help alleviate the complaints of millions of women. For that purpose, BPA had almost seemed like a home run—a *miracle* chemical—until researchers discovered a far more powerful synthetic estrogen called diethylstilbestrol, or DES. Around the same time, it was discovered that BPA had an alternate use with serious commercial potential. It could be used as the chemical backbone for an inexpensive material that was almost as hard as steel and as clear as glass: plastic. DES was brought to market as a drug, and BPA was routed to manufacturing instead.

In the decades that followed, these two chemicals went on to saturate our lives. DES was injected into millions of women, and plastics made with BPA exploded on the marketplace. We could suddenly fill our homes and lives with products of any size and type, and BPA made them inexpensive, easy to clean, shatterproof, and heat resistant. Susan Freinkel, in the book *Plastic: A Toxic Love Story*, paints it vividly: "In product after product, market after market, plastics challenged traditional materials and won, taking the place of steel in cars, paper and glass in packaging, and wood in furniture." But there was a problem.

Many compounds are brought to market only later to be revealed as damaging. Some of history's biggest fails include lead-based paints, asbestos building insulation, and partially hydrogenated fats. DES, the chemically similar sibling to BPA, had a comparable fate. "[It] initially appeared to be a benign and exciting reproductive technology, but in the long run [DES] had profound and damaging consequences for women," wrote sociology professor Susan Bell, in *Gendered Medical Science: Producing a Drug for Women*. For girls exposed while in their mothers' wombs, DES dramatically increased the risk of uterine malformations and rare vaginal cancers.

DES was finally banned from use in 1971, but BPA persevered. We now know that food and beverages stored in plastic made with BPA are able to leach this estrogenic compound. It's found in the dust created by our carpets, electronics, and furniture. And it commonly coats those heat-sensitive store register receipts, entering our bodies via our skin and hand-to-mouth behavior. For these reasons, 93 percent of people now have measurable amounts of BPA in their urine, with higher levels found in obese people.[2] The figures for phthalates are no more heartening.

While our exposure—our *dose*—is far lower than a syringe full of DES, BPA, like other endocrine disruptors, may be biologically active even at tiny doses. The FDA argues that BPA is safe, but the Endocrine Society, which publishes the leading peer-reviewed journals for hormone science around the world, disagrees, insisting that policymakers have overlooked, or altogether ignored, low dose toxicity effects.[3] It doesn't help that BPA-testing standards haven't been updated in over twenty years.[4]

One thing is clear: the safest level of BPA or phthalate exposure is none. Yet trying to avoid these chemicals completely is bound to be a frustrating (and futile) effort. The good news is that, thanks to our bodies' own detoxification pathways, these chemicals do not last very long once inside of us. Therefore, reducing your exposure to these fake estrogen compounds is likely to have a meaningful impact as your system of hormones and receptors recalibrates. Here are some guiding principles that should help:

▶ **Never microwave or reheat food in plastic.** Heat accelerates leaching of BPA and phthalates into your food, which is why you should never cook or store hot food in plastic. Always keep your plastic containers out of hot environments like your dishwasher, the sun, and your car.

▶ **Minimize consumption of foods and beverages sold in plastic containers.** Drinking out of a plastic bottle or cup won't kill you, but try to buy your liquids out of glass whenever possible. You don't know how that plastic container was stored before ending up in your hands. It could have been sitting in a truck's hot cargo bed for days, weeks, or even months!

▶ **Minimize consumption of canned foods and drinks.** The interior linings of cans are often made of BPA (yes, this includes canned beverages like sodas and seltzers). Removing all cans from your life won't be practical, but if you can at least cut down, you're ahead of the game. Acidic foods like tomatoes are especially likely to lead to leaching.

▶ **Avoid sous vide cooking.** This method involves cooking your food in a plastic bag placed in boiling water. Many restaurants keep food warm using this method. Remember that even BPA-free bags contain alternate plasticizing chemicals, and there is no reason to believe that they are safe.

▶ **Eat home more often.** As a result of food prep and storage, restaurants are a major source of phthalates and bisphenols. A study of over ten thousand people found that adults who had eaten the most food away from home had on average 35 percent higher levels of phthalates in their blood the next day.[5] Concentrations were higher (55 percent) for adolescents, probably due to greater fast food consumption.

▶ **Replace plastic storage containers with glass or ceramic.** Glass and ceramic not only can be used to cook but are easy to clean, are dishwasher safe, and look nicer. Don't worry about the lids unless they come in contact with your food.

▶ **Minimize use of plastic cutlery, plates, and cups.** Not only will the environment appreciate it, but it will reduce exposure to plasticizers like BPA, phthalates, and styrene (an endocrine disruptor and carcinogen).

▶ **Avoid fragranced products.** This includes most dish soaps, laundry detergents, fabric softeners and fresheners, deodorizers, and personal care products. Instead, look for fragrance-free products or products scented naturally with plant-based essential oils.

▶ **Toss old containers.** Plastics degrade with time, so if you've had plastic containers sitting in your cupboards for years and they're showing signs of wear and tear, it might serve you to toss them.

▶ **Skip the receipt.** Unless it's a major purchase, forgo the receipt. If you need it, wash your hands soon after. Always encourage children to do the same.

▶ **Plastic tea bag? Move oolong!** A Canadian research team found that steeping a single plastic tea bag released around 12 billion microplastic and 3 billion nanoplastic particles, yielding 16 micrograms of ingestible plastic per cup. Opt for paper tea bags or loose-leaf brewing methods instead.

INNOCENT UNTIL PROVEN GUILTY? NO THANKS.

Innocent until proven guilty: good for a justice system, bad when applied to newfangled chemicals to which humans and other animals are to be routinely exposed. Industrial creations, from food products and supplements to medicines and even medical devices, frequently infiltrate our lives before rigorous, long-term testing is ever done. These chemicals often escape the regulatory scrutiny of medicines or supplements simply because we don't ingest them. Other times, the complexity of our bodies doesn't allow us to see if a product is harmful until it's too late. Absence of evidence is not evidence of absence, and the newer the product, the higher the burden of proof it should bear before being fed, exposed to, or implanted into people. There are simply too many examples throughout human history where we were wrong.

A Nonstick Nightmare

Perfluorinated alkylated substances (PFASs) are also utterly ubiquitous throughout the modern world. PFASs help repel oil and water, so waterproof clothing, carpeting, upholstery, car parts, sealants, food wrapping papers, firefighting foams, and cookware all exploit the seemingly magical chemical properties of PFAS chemicals.

Unfortunately, these chemicals have been identified not just as potent endocrine disruptors, but as possible carcinogens. Animal studies have linked some PFASs with cancers including kidney, prostate, rectal, and testicular. They have also been linked to liver and thyroid problems and abnormal fetal development. Studies have shown that people exposed to higher levels of PFAS chemicals have higher total and LDL cholesterol, and they may make it more difficult to keep weight off after losing weight.[6] (Of course, these chemicals are commonly found in fast food and processed snacks, which can confound such findings.)

While some of the most well-studied PFAS chemicals have been banned from use by the FDA, 98 percent of all people still have detectable levels of PFASs in them, which means that we are likely still suffering their effects. Part of the reason for this is that PFAS chemicals linger in the body for years longer than phthalates and BPA.[7] And despite the ban, this hasn't kept manufacturers from finding chemically similar compounds to replace them with. Consumer awareness has led only to pacification as manufacturers hide newer (and often equally sketchy) chemicals from view.

Remember: it is nearly impossible to eliminate your exposure to these chemicals today, and chronically stressing about them is not a viable solution for your health. Therefore, cutting down on exposure is a far more attainable goal, and one that is less likely to drive you crazy in the process. Here are some tactics that will help:

▶ **Toss coated nonstick cookware.** The safest cookware to use is stainless steel (look for versions without nickel), cast iron, and ceramic. There are some nonstick pans on the market that claim to be PFAS-free, but the jury is still out on their safety.

▶ **Avoid smooth dental tape.** Dental *tape* is made using PFAS, which allows it to easily slide between teeth. New research suggests that the PFAS chemicals in the tape are not inert; they can enter circulation and cause health problems.[8] Use floss instead, which is also more effective at cleaning teeth due to its abrasive texture.

▶ **Avoid stain-resistant carpets, rugs, and furniture.** Though stain resistance can be useful, the PFAS particles in these products can easily become airborne and permeate our bodies. Young children in particular have very high concentrations of these and other chemicals due to their proximity to the floor and frequent hand-to-mouth behavior. Remember that children are particularly vulnerable to the effects of endocrine disruption.

▶ **Avoid foods wrapped in paper with a slick inner lining.** These linings keep the papers oil-proof and are commonly used to wrap burgers, burritos, and other convenience foods. Do not store or reheat your foods in these papers.

▶ **Avoid waterproof products unless you really need them.** Instead, look for coats, hats, boots, and tents labeled "water resistant," which are less likely to be treated with PFAS chemicals.

▶ **Use a reverse osmosis water filter.** PFAS chemicals pose a substantial environmental threat, having been identified in drinking water all over the United States. By separating water from its contaminants, a reverse osmosis water filter can remove up to 90 percent of a wide range of PFAS chemicals.

WHEN NOT TO USE A CAST-IRON PAN

Generally speaking, cast-iron pans are a great alternative to nonstick pans. Not only are they free of toxic chemicals, they can add significant amounts of iron to your food—but for some, that's a problem. Iron is essential, but if you ingest too much, it can build up in your blood and act as a pro-oxidant. This means that excessive iron levels can actually damage your organs and accelerate aging. Who is at risk? Men, postmenopausal women, and those with genes that increase iron absorption and/or storage—so-called *hereditary hemochromatosis*. If any of these apply to you, you should get all the iron you need from nutrient-dense foods like grass-fed beef or chicken and use your cast iron sparingly. (You can also just avoid using it to cook the foods that leach the most iron. Those would be acidic foods like tomato sauce and fatty foods like steak and eggs.) For *premenopausal* women, vegans, and vegetarians, or if you donate blood regularly, using a cast-iron pan can be a great way of increasing the iron content of your food.

Flame Retardants

Decades ago, house fires would claim thousands of lives every year, but the tragedy wasn't due to cooking mishaps or our furniture spontaneously combusting. Smoking in the house was common, and smoldering cigarette embers that landed on our sofas and easy chairs set whole homes ablaze. With pressure to find a solution, the tobacco industry pawned responsibility onto manufacturers of furniture, forcing them to incorporate the use of chemical flame retardants.[9] The result? Utter ubiquity of endocrine disruptors in the environment and in our bodies.

The specific chemicals in question are PBDEs, or polybrominated diphenyl ethers, which have now been in use for over thirty years in consumer electronics, furniture, and mattresses. They've been linked to cancer, neurological deficits, and impaired fertility. Peer-reviewed studies have shown that even a single dose of these chemicals administered to mice during development can cause permanent changes to the brain, leading to impairments in learning, memory, and behavior. In humans, studies have shown that higher levels of PBDEs in umbilical cord blood coincided with lower IQ in childhood, even after other complicating factors like the mother's IQ were controlled for.[10]

Like BPA, phthalates, and PFASs, PBDE chemicals were probably designed to stay put, but they easily migrate, and since they are most abundantly used in furniture, their most frequent hangout is house dust. As a result, we are frequently exposed to them, and the highest levels of contamination are found in those most vulnerable to their effects: children and infants. And let's not forget our pets, who love to roll around on our carpets and furniture—they are at high risk, too. A 2007 study found PBDE levels in cats twenty to one hundred times greater than median levels in US adults.

Cutting down on your exposure to these endocrine disruptors isn't easy, but it is indeed possible. Ensure that your home has adequately placed and fully operational smoke alarms, and follow these simple steps to purge the chemicals:

▶ **Choose flame retardant–free furniture.** Flame retardants are not necessary to reduce house fires. Many of the larger manufacturers are in the process of removing these toxic chemicals from their wares.

▶ **Have children? Read labels.** Clothing made for children often contains chemical flame retardants. Always buy children's clothing (especially pajamas) labeled as flame retardant–free.

▶ **Avoid farmed fish.** European and US farmed salmon have particularly high levels of PBDEs.[11] Choose wild fish (e.g., salmon) whenever possible.

▶ **Choose brominated flame retardant–free electronics.** Many companies are now forgoing brominated flame retardants (BFRs), which include PBDEs, in their products. The environmental organization Greenpeace releases an annual "Guide to Greener Electronics," which you can check for up-to-date information on companies making the greatest effort.

▶ **"Wet dust" or use a HEPA filter vacuum.** Dusting with a damp cloth or sponge helps to trap contaminated dust, but using a vacuum with a high-efficiency particulate absorbing (HEPA) filter is the ideal way to remove dust if you have conventional furniture, carpets, and curtains.

▶ **Use an air filter.** A high-quality air filter can help cut down on household dust, which will reduce your exposure not just to flame-retardant chemicals, but many of the previously mentioned chemicals as well.

IS YOUR BUILDING MAKING YOU SICK?

Air quality in buildings is a growing concern for public health officials, especially as they become more tightly sealed to save on heating and air-conditioning costs. People subjected to poor indoor air quality display reduced cognitive performance and even complain of various symptoms that are the basis of a real medical diagnosis, "sick building syndrome" (SBS). Sufferers report fatigue, headache, dizziness, and nausea, and the severity and duration of symptoms are directly tied to time spent in the building.[12]

SBS is thought to be a direct result of exposure to indoor pollutants, including many of the compounds I've already mentioned, plus carbon dioxide (which in higher concentrations may affect cognitive function) and about 150 volatile organic compounds that we naturally excrete, including carbon monoxide, hydrogen, methane, ammonia, and hydrogen sulfide. The most common indoor air pollutant? Formaldehyde, which is released via pressed wood products and countless other consumer products.

The solution to building sickness? Try to encourage ventilation in your home or office with either open windows or vents. Wet dust and use a HEPA vacuum regularly. And, if you spend many hours in one location, consider an air purifier, or even better (and much less expensive), mother nature's own air purifiers: plants. On page 167 I will offer some suggestions for specific plants as well as placement.

Bathroom Deeds

We interact with many compounds on a daily basis that can lead to unpredictable health consequences. Some, like those mentioned above, are ingested unknowingly in the countless consumer products leaching them into our air and food. Others we ingest willingly, often without knowing the full breadth of their health implications. The following section will home in on the suspicious chemicals found in our medicine cabinets.

Cosmetic Carnage

Parabens function as a preservative and are used to prevent the growth of microorganisms. You can tell when a product has parabens in it by looking at the ingredient list; the compounds in question end in "-paraben." (Common examples include methylparaben and propylparaben.) They are often found in shampoos, face washes, deodorants, personal lubricants, and lotions, and are also common in packaged foods where they help to extend shelf life.

Parabens are easily absorbed into the body either through the mouth or on the skin and possess hormone-disrupting abilities. They've caused cancer in lab animals, and while such a causal connection has not been established in humans, they have indeed been *linked* to various human cancers. For example, parabens have been discovered in tumors of the breast. This doesn't prove that they caused the tumor, but given their hormone-altering potential, their presence is certainly a suspicious one.[13]

Parabens that we eat are expected to be "detoxed" by our liver and kidneys, but parabens absorbed in the skin may accumulate. Daily slathering of creams containing parabens therefore provides a reasonably worrisome door for chronic exposure, compounded by what we ingest orally. The good news is, as with some of the other endocrine-disrupting chemicals, they don't stick around long once you stop using these products.

Thankfully, there is no shortage of healthier products on the market, now paraben free. Support local health stores and smaller brands by looking for paraben-free products or, better yet, make your own versions. One tactic that may be a safe bet: don't put anything on your skin that you wouldn't eat!

NSAIDs

Any time we try to tinker with our bodies, some unintended effect—however slight—is bound to occur. This is particularly true for nonsteroidal anti-inflammatory drugs, or NSAIDs. This category of drugs includes many common pain relievers, like aspirin, ibuprofen, and naproxen. Since these drugs are available without a prescription, many assume that they are harmless.

One risk of regular NSAID use is cardiovascular events like heart attack. Though the mechanism is complex, one possible source of harm is to the mitochondria of heart cells, reducing their ability to produce energy. This increases the production of reactive oxygen species, aka free radicals, which can create damage in the heart tissue. These drugs can also cross the blood-brain barrier, and it's as yet unknown whether they can affect the mitochondria of brain cells. Nonetheless, those of us at risk for neurodegenerative conditions should exercise caution with routine use, as brain mitochondrial dysfunction is associated with cognitive decline. (No clear link has thus far been established for dementia and NSAID use.)

NSAIDs also have a negative effect on the health of your gut, as they are able to both irritate and block the enzymes that protect its inner lining. In fact, gastrointestinal side effects like ulcers and bleeding are common with routine use. Making matters worse, NSAIDs may also tinker with the beneficial bacteria that live within, altering your microbial colony in a way that makes you vulnerable to opportunistic pathogens like *Clostridium difficile*.[14] *C. difficile* infection (and subsequent diarrhea) causes half a million hospitalizations and thirty thousand deaths annually, according to the CDC.

Finally, these drugs indiscriminately alter your body's natural inflammation pathways. Inflammation gets a bad rap, partly because of its involvement in pain and swelling. But it is also how we reap the benefits of exercise or sauna. A good workout or sweat, for example, causes a temporary spike in inflammatory markers, which causes your body to respond in a positive way. Studies show that broad-sweeping anti-inflammatory drugs like NSAIDs can inhibit the beneficial effects of exercise, including muscle growth.[15]

Remember: just because a drug is available over the counter does not make it unconditionally safe. Try to avoid persistent, daily use of NSAIDs, saving them for more serious pain instead. For milder aches and pains, try curcumin (a component of the turmeric root) or the omega-3 fat EPA; both are anti-inflammatory and do not carry the NSAID-associated risks.

Acetaminophen

Acetaminophen is another common over-the-counter pain reliever. Unlike NSAIDs, acetaminophen does not block the inflammation that is causing pain. Though the exact mechanism of action is controversial, it is thought to work on the nervous system, raising one's pain threshold to make pain more tolerable. But pain isn't the only feeling the drug seems to numb: researchers from Ohio State University gave students 1000 mg of acetaminophen and found that compared to placebo, they exhibited less empathy and blunted positive emotion when shown depictions of people experiencing either pleasure or pain.[16]

Acetaminophen may also impact the brains of your children-to-be. Persistent prenatal exposure to acetaminophen has been linked to increased autism spectrum and hyperactivity symptoms in children,[17] and frequent use increased the odds of language delays in two-year-old girls by more than sixfold.[18] And while the mental effects of acetaminophen are still being elucidated, its effects elsewhere in the body are more certain.

Acetaminophen is well known to cause a sharp reduction in the liver's ability to create glutathione, an important antioxidant and master detoxifier for your body as well as your brain.[19] Just how much is needed to cause permanent liver damage or even death? The margin of safety is narrow—just a few doses more than what you might take for a pain condition—making acetaminophen overdose a major cause of emergency room visits in the Western world. (The antidote to acetaminophen overdose is actually a glutathione precursor called N-acetyl cysteine.)

As with NSAIDs, occasional use won't hurt you, and certainly, if you need it, by all means, use it! But please be mindful of chronic use, particularly if pregnant.

Anticholinergic Drugs

Insomnia, allergies, anxiety, and motion sickness—what do they all have in common? They are all frequently treated with a class of drugs known as anticholinergics. Some of these drugs are available only with a prescription, but many are found over the counter, which is surprising given their ability to screw with your brain.

Anticholinergic drugs work by blocking a chemical called *acetylcholine*. Below the neck, acetylcholine is responsible for involuntary muscular contractions. This is why an anticholinergic may be prescribed to help quiet an overactive bladder. But in the brain, acetylcholine is critical for learning and memory, and regular use of an anticholinergic drug can cause cognitive problems in as little as sixty days.[20] For longer-term users (three years or more), there is a sharply increased risk of developing dementia—by up to 54 percent.[21]

Once in a while use may be okay, but keep in mind that even occasional use of a strong anticholinergic can cause acute toxicity. A mnemonic is often taught to med school students to help them remember the symptoms:

Blind as a bat (dilated pupils), red as a beet (flushing), hot as a hare (fever), dry as a bone (dry skin), mad as a hatter (confusion and short-term memory loss), bloated as a toad (urinary retention), and the heart runs alone (rapid heartbeat).

Take note of this abridged list of the most common anticholinergic medications to avoid. If you are currently taking any on doctor's orders, have a conversation with them about safer alternatives:

VERY COMMON ANTICHOLINERGIC DRUGS

Dimenhydrinate	Motion sickness	Strong anticholinergic
Diphenhydramine	Antihistamine/sleep aid	Strong anticholinergic
Doxylamine	Antihistamine/sleep aid	Strong anticholinergic
Oxybutynin	Overactive bladder	Strong anticholinergic
Paroxetine	Antidepressant	Strong anticholinergic
Quetiapine	Antidepressant	Strong anticholinergic
Cyclobenzaprine	Muscle relaxant	Moderate anticholinergic
Alprazolam	Antianxiety	Possible anticholinergic
Aripiprazole	Antidepressant	Possible anticholinergic
Cetirizine	Antihistamine	Possible anticholinergic
Loratadine	Antihistamine	Possible anticholinergic
Ranitidine	Antiheartburn	Possible anticholinergic

Aluminum

Who doesn't have fond memories of wrapping corn and other veggies in tinfoil and throwing them on a summer barbecue? Having partly grown up on the eastern tip of Long Island, I remember how my mom would relish buying locally grown sweet corn every summer and doing exactly that for my family. What we (and countless other families) didn't know is that tinfoil is not inert—it's able to leach aluminum into the foods wrapped in it.

Aluminum serves no purpose in the body and is not naturally found within us, but thanks to the small amounts now found in produce, meat, fish, and dairy, in our water supply, and in our cosmetics (which we readily absorb through our skin), aluminum is now detectable in nearly all humans. It is generally considered safe, but because the modern world has increased our exposure to the metal, it has rightfully received scrutiny by scientists and wellness warriors alike.

The data on aluminum is conflicting. Some studies have connected it to various forms of cancer and other studies (possibly not large or lengthy enough) have found no such association. When it comes to dementia, the evidence is somewhat more concerning. In a 2016 meta-analysis of environmental exposures, aluminum was the only heavy metal to have garnered high-quality studies suggestive of a link. Worth noting, it's been found in the brains of patients with Alzheimer's disease, but even so, its presence may be an effect of the disease, rather than the cause.

When it comes to aluminum, the less of it there is lingering in our bodies, the better. To *minimize* unnecessary exposure, and give your body a chance to eliminate the rest, consider these simple principles:

- ▶ **Ditch the antiperspirant.** Use a natural underarm deodorant instead. Or consider forgoing such products entirely.
- ▶ **Avoid cooking with tinfoil.** Tinfoil is not considered dangerous, but minimizing use is probably worthwhile, especially when cooking meats and acidic foods and at high temperatures, which accelerates the leaching.[22] Use glass or stainless steel instead.
- ▶ **Use a water purifier that reduces aluminum content.** Reverse osmosis purifiers, for example, are able to remove significant aluminum from drinking water.
- ▶ **Avoid regular antacid use.** Never mind that antacids block stomach acid, which is required for the absorption of numerous vitamins and minerals (more on this shortly); many antacids are aluminum-based, providing a very high dose of the nonessential metal. If you are required to use an antacid for medical reasons, look for versions without aluminum.

▶ **Sweat.** Sweat is a major route of excretion for aluminum.[23] Enjoy vigorous workouts or time in the sauna, both of which will help purge a significant quantity of aluminum.

Antibiotics

We've all taken antibiotics at one point or another. Undeniably, antibiotics can and have saved lives. What may shock you, however, is that 30 percent of antibiotics prescribed in the United States are thought to be completely unnecessary.[24] The consequences may be profound.

In your gut, there are good bacteria and there are bad bacteria. The good guys help to keep the bad guys (i.e., potential pathogens) in check, while also churning out powerful, health-promotive compounds like anti-inflammatory fats and vitamins. Some strains of bacteria, such as *Lactobacillus rhamnosus*, can even reduce intestinal absorption of various harmful heavy metals like mercury and arsenic, two neurotoxins with increasing concentration in our environment.[25]

When we take antibiotic drugs, it's the equivalent of dropping a nuclear bomb on the gut; they devastate indiscriminately, killing off many of our beneficial bacteria. Though the bacteria mostly return over time, one study found the intestinal ecosystem was still catching up six months after a high-dose course of the commonly prescribed antibiotic ciprofloxacin (Cipro).[26] Aside from losing out on health-promotive activities of "good" bacteria, such destabilization can allow bacteria that can make you sick, such as *C. difficile* (mentioned in the NSAIDs section earlier), to multiply and take over.

Taking a pill is one thing, but if you regularly eat factory-farmed meat or farmed fish, you're also ingesting antibiotics in your food. Low doses of antibiotics are routinely given to farm animals to fatten them up, and because the animals are typically confined to horrible living conditions, they are often also medicated prophylactically to prevent them from getting sick. Residues of these antibiotics persist in their meat and milk.[27] It's no wonder our waistlines are growing; low-dose antibiotics may be fattening us up as well.

Here are some ways to be kind to your beneficial bacteria:

▶ **Avoid broad-spectrum antibiotics if possible.** Always consult with your doctor to see whether your antibiotics are really necessary.

▶ **Eat organic produce.** Pesticides, herbicides, and fungicides are all antimicrobial and have the potential to disrupt the health of your intestinal immigrant population (more on pesticides shortly). Organic produce has far lower levels of pesticides.[28]

▶ **Choose grass-fed, pasture-raised, and antibiotic-free meats.** Remember: you aren't just what you eat; you are what what you eat eats.

▶ **Reconsider probiotics with those antibiotics.** New research suggests that probiotics taken after antibiotics can delay your ecosystem to returning to its pre-antibiotic state.[29] A better strategy is likely to consume an organic, veggie-rich diet to encourage repopulation naturally.

▶ **Focus on fiber.** By consuming a diverse array of plant fibers, you encourage the proliferation of the countless beneficial strains of bacteria in your gut. Incorporate dark leafy greens, alliums (garlic and onions, for example), cruciferous vegetables, roots and tubers, and fruit. I'll give you a robust shopping list including many of these items in chapter 7.

A SLOW CHEMICAL FATTENING?

We're often told that calorie excess (and moral weakness) is the sole determinant of weight gain, but the amount of fat you store may be at least partly determined by the environmental chemicals that you are exposed to. This was a suggestion set forth by a large, multigenerational study from York University that found that for the same amount of calories consumed and energy burned, a person in 2006 would weigh 10 percent more than a person in 1988.[30] In other words, someone today would need to eat less, and work out more frequently, to achieve the same weight as someone fewer than twenty years ago. Though the observational study couldn't rule out other variables, low-level antibiotics and hormone-disrupting chemicals may be altering our ability to stay lean and healthy, even despite our best efforts.

Fluoride

Caring for your teeth is important. A great smile is how you present to the world, and tooth decay is correlated to worse cardiovascular and neurological health. The discovery of fluoride has been instrumental in helping tooth decay go from a global epidemic to a comparatively rare phenomena, and fluoride is now added to commercial toothpastes and municipal drinking water as a result.

There's no doubt that fluoride can help prevent tooth decay for those with developing teeth (i.e., children) and for people at otherwise high risk. But fluoride is also a potential endocrine disruptor, and you likely do not need any additional fluoride if you do not regularly eat foods that damage your enamel. Tooth decay begins to occur when the bacteria that live in your mouth ferment starch and sugar (especially from refined grain products) and secrete harmful acids that burn through your tooth enamel as a result. Avoiding sugar and grain products combined with regular oral care can reduce your need for fluoride.

Here are some guiding principles that you can use to improve your oral (and overall) health, sans potentially health-disrupting chemicals:

▶ **Use fluoride-free toothpaste.** Or simply make your own—combine coconut oil, baking soda, a pinch of sea salt, and some xylitol or erythritol and cinnamon for an effective paste. Even simpler, try brushing with activated charcoal; it's a powerful whitener, believe it or not!

▶ **Floss regularly.** Pick dental floss instead of tape, since the latter is made with PFAS chemicals, as mentioned on page 140. At a minimum, floss every night just prior to brushing.

▶ **Ditch antiseptic mouthwashes.** Alcohol-based mouthwashes indiscriminately kill both good and bad oral bacteria. As with your gut, good oral bacteria help police the bad bacteria.

▶ **Avoid processed grains, refined flours, and sugar.** Microbes love to ferment these easy-to-break-down foods, which enables them to excrete damaging acids from the cushy biofilm that they form on your teeth.

▶ **Avoid strong acids.** Lemon in your water or tea is fine, but avoid sucking on the fruit, as the acid can damage tooth enamel. Same goes for undiluted vinegar.

▶ **Get your vitamins D, A, and K_2.** Weston A. Price was a dentist who championed the merits of nutrition for dental health. Vitamins D and A work synergistically with vitamin K_2 (found in grass-fed beef, dairy, and natto) to direct bodily supplies of calcium toward building healthier bones and teeth.

Acid-Blocking Medications

Millions of Americans take acid-blocking medications. Why wouldn't they? Larry the Cable Guy offers them as a heartburn-free pass to enjoy all the pizza, chili dogs, and onion rings one could want! But is the acid our stomach secretes the bad guy, or is its overproduction merely another signal that we are eating the wrong foods?

You need stomach acid to properly break down and absorb nutrients from food. Folate and B_{12}, two nutrients that are essential for proper cellular detoxification and brain function, both require stomach acid for absorption, as do minerals such as calcium, magnesium, potassium, zinc, and iron.[31] Stomach acid also helps us break down the protein in our food and helps prevent allergies associated with incomplete digestion of protein—a major challenge for a portion of the population.

The second advantage of having healthy levels of stomach acid is that it keeps the upper gastrointestinal tract free of bacterial overpopulation. Bacteria don't like stomach acid—this is why you don't have a lot of bacteria in your stomach. But low stomach acid (or *hypochlorhydria*) can be a risk factor for small intestinal bacterial overgrowth, or SIBO. This is an uncomfortable condition in which bacteria can set up camp too high in the digestive tract (far from its usual hangout, the colon), impairing nutrient absorption and creating uncomfortable symptoms such as gas (which is not usually produced there) and diarrhea.

How might the rest of us better manage our stomach acid? Here are some tips:

▶ **Lose weight.** Being overweight is a major risk factor for reflux, partly by putting pressure on the junction between your stomach and esophagus.

▶ **Cut out the refined carbohydrates and reduce carbohydrates in general.** Studies suggest that low-carbohydrate diets can improve symptoms of reflux. Bid adieu to grain-based pastas, bagels, cereals, wraps, and rolls.

▶ **Don't eat for two to three hours before bed.** This is in line with the recommendations made in chapter 2, which will also help preserve your circadian alignment.

Chemical Sunblocks

Being fair-skinned, my mother was always afraid of the sun, and family trips to tropical locales always meant bringing with us broad-brimmed hats and heaps of drugstore sunblock. But the chemicals our families have been advised to lean on by our most trusted health professionals to prevent skin cancer may not only be ineffective at doing so, they may comprise some of the most dangerous chemicals unleashed on our environment.[32]

One of the primary dangers of common chemical sunblocks is that they are able to mutate into very harmful compounds. This was recently demonstrated with avobenzone, a sunblock found in nearly all drugstore brands when subjected to a combination of ultraviolet radiation from the sun and chlorinated pool water. The chemicals that avobenzone transformed into are known to cause liver and kidney problems, nervous system disorders, and cancer, and this occurred *right* on the skin of human subjects in the experiment.[33]

Avobenzone isn't our only worry; it is usually found alongside other chemicals that come with other problems. Oxybenzone, for example, has been found to exhibit endocrine-disrupting potential. This means that, like the phthalates, BPA, nonstick chemicals, and parabens mentioned earlier, it may be able to affect growth, development, and reproduction. This is worrisome, as these sunblocks have already been found polluting our children, in amniotic fluid and even breast milk, to say nothing of the fact that we slather them in it every summer.

Similar to parabens, avobenzone and oxybenzone are easily able to go from our skin to our blood. In an article published in the *Journal of the American Medical Association*, researchers found that after people applied the sunblock all over their bodies and then reapplied according to directions, they had enormous increases of these compounds in their blood.[34] In fact, they reached a blood concentration far higher than the Food and Drug Administration's Threshold of Toxicological Concern, which is the ceiling for which a chemical is presumed safe by the organization. Since this startling finding, the FDA has sent a mandate to manufacturers to provide proof that these chemicals are actually safe, begging the question: why hadn't this happened earlier?

It should be clear by now that the sun is your friend. We need it to synthesize vitamin D, set our bodies' internal clocks, and even help our blood vessels function more effectively. But burning is no good, either; each of us has a responsibility to take what we need from the sun, and then protect ourselves from sun damage. The solution is to avoid needing to use sunblock by avoiding *excessive* sun exposure (revisit page 81 to see how much time in the sun you actually need to synthesize vitamin D). Here are some other tips that may help:

▶ **Use a mineral-based sunscreen.** If you do require sunblock for a long day in the sun, opt for a safe zinc oxide–based sunscreen that forms a physical barrier (i.e., not a chemical one) between your skin and the sun.

▶ **Avoid chemical sunblocks.** These include avobenzone, oxybenzone, octocrylene, and ecamsule. They are found in nearly all common brands of sunblock as well as various lip balms and lipsticks. Simply check the active ingredients to determine.

▶ **Take an astaxanthin supplement.** Astaxanthin is the deep orange pigment found in shrimp, salmon, and salmon roe. There is preliminary evidence from rodent studies and human trials that astaxanthin may help to reduce damage from the sun's UV rays.[35] Start with 4 mg per day, but if you are fair-skinned or prone to sunburn, you may take up to 12 mg per day.

In the Kitchen

Many of the chemicals discussed above are used to produce the products that define modern life. The unfortunate side effect is that many of them end up in our food, where we ingest them to myriad consequences. In the following section, I will focus on toxins that are commonly found in our food as well as our environment, along with practical ways that you can minimize your exposure.

Mercury

Mercury, like aluminum, is a heavy metal that is common in the natural environment. It has no function within the body and can become toxic at high levels. It is found in some factory-farmed live-stock, vegetables and grains grown in mercury-contaminated soil, and most concentrated in certain fish.[36]

If you are regularly exposed to mercury in your environment—particularly inorganic mercury, the most dangerous type—this is bad news. That's generally only a concern if you work with mercury as part of your job. Eating commonly available fish, which contains a form of mercury created by bacteria, may not be cause for panic. That's because even though this form of mercury can be toxic at high levels, many fish are rich in the mineral selenium, which may "block" some of mercury's toxic effects.[37]

To stay healthy, your body and brain depend on a vast network of antioxidants, many of which rely on selenium. Mercury can bind to these important, selenium-based protector chemicals, rendering them incapacitated and allowing damaging processes, like oxidation, to run amok. When oxidation occurs too fast for the body to repair itself, aging and disease ensue, and the brain is particularly vulner-able. But with ample selenium, the small amounts of mercury in an otherwise healthy fish may be harmless.

The key to safe fish consumption therefore may lie in there being a high selenium-to-mercury ratio in the fish. And while this hy-pothesis needs further testing, it is noteworthy that original human studies linking seafood consumption to mercury toxicity involved the pilot whale—a mammal, not a fish, and one whose flesh contains far higher mercury content than selenium.[38]

When it comes to our most commonly consumed fish, the benefits outweigh the risks—just be sure that your fish is either baked, broiled, or grilled as opposed to fried, which increases the amount of unhealthy oils in your diet. For older adults, eating sea-food twice a week protected their cognition over time compared to

Fish with High Selenium-to-Mercury Ratio (Safe)	Fish with More Mercury Than Selenium (Avoid These)
Albacore*	Shark
Herring	Swordfish
King mackerel*	Whale (a mammal)
Mackerel	
Salmon	
Sardines	
Tuna*	

*These fish have the highest levels of mercury and should be moderated in children and pregnant or lactating women.

nonconsumers, and the effect was even stronger for carriers of the common Alzheimer's risk gene, ApoE4.[39] An earlier study from the same group found that fish consumption was related to reduced Alzheimer's-related brain changes, and that brain mercury levels (though correlated to fish consumption) did not increase Alzheimer's pathology.[40]

Eating fish imparts benefits when young, too. Prenatal fish consumption may lead to stronger brain development for the baby, and children who ate fish at least once a week slept better and had IQ scores that were four points higher, on average, than those who consumed fish less frequently or not at all.[41] Fifteen-year-old males who ate fish at least once a week showed a 6 percent increase in intelligence scores by the age of eighteen. Those who ate it more than once a week almost doubled that improvement.[42]

To ensure low exposure to mercury while enjoying all the benefits of fish consumption, abide by these simple principles:

▶ **Sweat.** Sweating is a long-standing heavy metal detoxification method.[43] Engage in exercise or hit the sauna (with your doctor's permission, of course), and know in doing so that you are helping your body rid itself of significant quantities of mercury.

▶ **Choose oily fish.** Tuna may be more common, but Alaskan wild salmon, sardines, and Atlantic mackerel are not only are low in mercury, they're also healthier due to the abundance of omega-3 fats they contain. And they're environmentally sustainable.

▶ **Include higher selenium, nonfish foods in your diet.** Pork, turkey, chicken, eggs, beef, sunflower seeds, mushrooms, and especially Brazil nuts all contain significant quantities of selenium.

▶ **Opt for free-range poultry and grass-fed beef.** Animals allowed to graze or forage are less likely to contain mercury because they are not fed processed fish meal.

▶ **Consider replacing old mercury dental fillings.** These fillings increase the burden of mercury in the blood, urine, and brain.[44] To what consequence? That's a subject of debate. If you do choose to get them replaced, find a dentist who can minimize collateral exposure. And ensure that the replacement material is not BPA-based, which, unfortunately, is more common in dental fillings than you'd think.

Arsenic

Arsenic poisoning can lead to skin problems, nausea, vomiting, and diarrhea, and in high enough doses can cause cancer, heart disease, and death. If that's not enough, it is also an endocrine disruptor, and can interfere with the glucocorticoid system (you may have heard of cortisol, one of its primary hormones).[45] Disrupting this system has been linked to fat gain, muscle loss, suppression of the immune system, insulin resistance, and high blood pressure, among other things.

Recent research has found that inorganic arsenic is able to concentrate in the hull of rice. For this reason, I recommend that children and pregnant women avoid rice completely and suggest moderation or avoidance for the rest of us (the good news is that rice is almost pure starch, containing little of anything else, nutrient-wise). Here are some other tips to help you avoid unwittingly feeding yourself and your loved ones arsenic:

▶ **Avoid brown rice grown in the southern United States.** Opt instead for rice grown in California, India, or Pakistan, which has far less arsenic in it, according to *Consumer Reports*.

▶ **Avoid products made with rice flour.** These often contain the highest levels of arsenic. They include rice crackers, rice cereals, rice milk, and protein powders made with rice.

▶ **Use a water purifier.** A reverse osmosis water purifier can reduce the arsenic content of your tap water.

▶ **Eat white rice instead.** When unsure of where the rice has come from—in restaurants, for example—order white rice.

▶ **Choose grains that are less likely to be arsenic-contaminated.** While I recommend limited (if any) grain consumption, should you choose to consume grains on occasion, quinoa can be a good choice, as it has much lower levels of arsenic than any of the types of rice tested by *Consumer Reports*.

Pesticides, Herbicides, and Fungicides

Did you know that a conventionally grown strawberry contains an average of eight different pesticides, according to research from the U.S. Department of Agriculture? Our food supply has become awash in various pesticides, herbicides, and fungicides meant to increase yield and, ultimately, profit.

Of those currently in use around the world, glyphosate is the most common.[46] Glyphosate is sprayed on crops to prevent weeds from attacking them, and it is also used to dry them out just prior to harvesting. More than thirteen billion pounds of glyphosate-based herbicides have doused the world's crops over the last decade. In the United States alone, its usage has increased nearly sixteenfold between 1992 and 2009. It has since been detected in our food, our water, the dust we inhale, and in us.

Some crops—namely corn, soy, and rapeseed, which is used to make canola oil—have been genetically engineered (genetically modified organisms, or GMOs) specifically for this reason: to withstand heavy spraying. Then we either eat these foods, or they are used to feed fish and livestock. This has led to significant glyphosate residue in all variety of foods, including fish, meat, berries, vegetables, baby formula, and grains.[47] (Think cooking makes you safe? It doesn't; residues persist long after heating.)

Generally speaking, pesticides, herbicides, and fungicides are neurotoxic and have potential endocrine disrupting abilities.[48] The question is, of course, at what dose? As discussed, this is difficult to answer. While a massive and concentrated exposure would be required to produce acute effects in people (including certain cancers), low-level exposures have been associated with various health issues, including lower cognitive test scores and behavioral and attention problems in children, asthma, and impacts on the reproductive and endocrine systems.

The cancer risks of persistent, low-level glyphosate exposure are under constant debate. The U.S. Environmental Protection Agency says that the herbicide "is not likely to be carcinogenic to humans," yet the World Health Organization's International Agency for Research on Cancer classified it as "probably carcinogenic to humans." And while the European Food Safety Authority sides with the EPA on glyphosate, some pesticides that are used in the United States are banned in Europe for their toxicity.

Organic isn't perfect, and it isn't always pesticide-free (there are plant-derived pesticides that are approved for use in organic farming systems, and processing itself can lead to cross-contamination of synthetic pesticides on organic produce). Still, meta-analyses have shown that organic has lower levels of pesticides and herbicides and higher levels of certain antioxidants. Switching to an organic diet not only helps our bodies purge these chemicals,[49] but if sustained, it may even help prevent several types of cancer, as was shown in an observational study of nearly seventy thousand adults.[50] The risk reduction for non-Hodgkin's lymphoma, the cancer most associated with the herbicides and pesticides used in conventional agriculture, was a staggering 86 percent.

Choosing organic means voting for a food production system that is always improving, better for the environment, and likely better for your health as well.

FAQ

Q: I'm living on a low income. I can't afford to buy everything organic. What do I do?

A: Not everything you buy must be organic. Here is a simple rule of thumb: if you eat the whole fruit or vegetable, try to buy organic. For example, bell peppers, cruciferous vegetables, berries, and leafy greens like spinach. Conventional (i.e., nonorganic) spinach had more pesticide residues by weight than all other produce, with an average of 7.1 different pesticides on every sample collected in 2016 according to the Environmental Working Group (one was a neurotoxic pesticide that is banned from use on food crops in Europe). If there is skin or a peel, it's safe to buy conventional. There is no need to purchase organic bananas, avocados, melons, or citrus, but if you plan to use or eat the peel (to zest, for example), buy organic. And if organic is totally off-limits, either due to budget or accessibility, fear not. Revisit page 37 for a rinse technique for added safety.

Cadmium

Cadmium is another heavy metal that is found in soil due to natural activities like volcanic activity, forest fires, and the weathering of rocks, but can also be added via industrial processes (coal burning, for example). When present in soil, cadmium concentrates in crops like cacao, dark leafy greens, tubers, and grain products like rice and bread.

Not only is cadmium toxic to the cardiovascular system, it is a known carcinogen linked to tumors of the kidney, lung, and prostate. It can cause kidney and liver damage, and it can also "infect" the brain, where it can cause cognitive problems. A 2012 study lead by Harvard researchers found that among children with the highest levels of cadmium, the odds of having a learning disability were over threefold higher compared to children with the lowest exposures.[51]

Here are some key ways to ensure the least possible cadmium exposure:

▶ **Get to know where your food comes from.** Crops grown in polluted areas are major dietary sources of cadmium, so if "local" means "off the local freeway," find a better source of chard.

▶ **Shop organic.** Organic produce has lower levels of cadmium by about half, according to a meta-analysis of over three hundred studies published in the *British Journal of Nutrition*.[52]

▶ **Become a chocolate snob.** The independent testing site Consumer Labs found that dark chocolate bars generally had low levels of cadmium (falling within the World Health Organization's limits of safety), but many cocoa *powders* were over the limit—some containing as much as five times over! (Cocoa *extracts* were A-OK.)

▶ **Avoid wheat dishes and bread.** Commercially prepared breads and wheat dishes are a top source of cadmium intake in the American diet.[53]

Lead

Lead, like aluminum, cadmium, and arsenic, is a metal that occurs naturally, but its concentration in our environment has steadily increased over the past century. Prior to 1978, lead was used routinely as an ingredient in paint. As with BPA and phthalates, its use made sense at the time; lead sped up drying, made paint more durable, and made it resistant to both moisture and temperature. But lead is a powerful neurotoxin, and it, too, has found its way into our bodies.

When lead-containing paint dries, it can chip off. This allows it to end up in places we least expect, such as in the soil near our homes. Children at play can easily get it on their hands (or pets on their paws), ultimately finding its way to their mouths. More insidious, however, is when lead-based paint has been used in high-use parts of the home, such as windowsills, door frames, stair banisters, or cabinetry. The wear and tear can create dust, which can end up on our skin, in our food, or even in our lungs.

Walls aren't the only place lead-based paint has been used, of course. Toys and furniture can all present opportunities for lead exposure. While antiques are the most obvious potential culprit, as recently as 2007, a major toy manufacturer (Fisher-Price) recalled about one million toys because they had been painted with lead-containing paint. Be careful of painted products that are made in China (which is where these products were made), especially for toddlers who may put them in their mouths. Though Chinese regulations are in place, a recent study of decorative paints sold there found that more than half exceeded lead regulations, with one third containing "dangerously" high levels of lead.[54]

Food grown in contaminated soil is another major source of lead, as are packaged, processed foods that come into contact with machinery that could leach the heavy metal. After reading chapter 1, it should be clear to you that processed food consumption comes with a number of potential downsides—add the possibility of lead contamination to the list. Particularly worrisome is the contamination of certain baby foods,[55] since lead is dangerous to the brain development of children (even very low blood lead levels can cause behavioral problems and lower IQ[56]). Certain fruit juices (apple and grape in particular) were also determined to have high levels of lead.

The ideal level of lead ingestion is none, as there is no established "safe" blood level. To ensure you and your loved ones are protected, follow these guidelines:

- ▶ **Avoid buying painted objects from China, especially for toddlers.** While many products can be safe, there is still a risk of poor quality control and the contamination of painted items with lead.
- ▶ **Test your water for lead.** Find contact information for your local water-testing center by calling EPA's Safe Drinking Water Hotline at 800-426-4791.
- ▶ **Check your home for lead-based paint.** If your home was built before 1978, there's a good chance that it contains lead-based paint. It's less of a problem if the paint is in good condition, but be sure to eliminate any dust or chipped pieces and throw them in the trash.
- ▶ **Wet dust and vacuum regularly.** If you suspect there is lead-based paint in your home, use a damp disposable cloth to dust, then throw out that cloth.
- ▶ **Avoid processed baby foods and fruit juices.** According to the FDA's Total Diet Study, the most likely to be contaminated were arrowroot cookies and "baby food" sweet potatoes and carrots. The juices most likely to be contaminated were apple and grape.
- ▶ **Consume lots of veggies.** Vegetables are particularly rich in minerals, which can reduce intestinal absorption of heavy metals like lead.

"Detoxing" the Modern World

If you follow the recommendations above, you should be able to minimize your exposure to many of the aforementioned toxins. But what of the toxins you've already accumulated? For these, we must detox. Unfortunately, the word *detox* has been co-opted by the wellness industry, which for all of its positive aspects, is financially fueled—at least in part—by selling you on the notion that you are inadequate. But you aren't inadequate. Your body actively detoxifies itself, and there are powerful dietary and lifestyle tools that you can use to support (and even boost) this process.

Keep Plants Around You

Plants don't just make our spaces more inviting; they clean the air. The air-purifying ability of plants was originally studied for their utility on space stations where they could help produce fresh, clean air for our astronauts. In his excellent book *How to Grow Fresh Air*, NASA research scientist Dr. Bill Wolverton details the plants that are most effective at removing various chemical vapors. According to his studies, these are the top ten plants for cleaning the air in your home or office, rated for their efficacy and ease of maintenance:

1. Areca palm (*Chrysalidocarpus lutescens*—recently changed to *Dypsis lutescens*)
2. Lady palm (*Rhapis excelsa*)
3. Bamboo palm (*Chamaedorea seifrizii*)
4. Rubber plant (*Ficus robusta*)
5. Dracaena Janet Craig (*Dracaena deremensis* "Janet Craig")
6. English ivy (*Hedera helix*)
7. Dwarf date palm (*Phoenix roebelenii*)
8. Ficus alii (*Ficus binnendijkii* "Alii")
9. Boston fern (*Nephrolepis exaltata* "Bostoniensis")
10. Peace lily (*Spathiphyllum sp.*)

Keep plants in your home, and if you spend extended bouts of time in any one location, make sure that you have plants in your "personal breathing zone," defined by Dr. Wolverton as six to eight cubic feet around you. And note that some of the above plants are toxic if ingested by dogs or cats, so inquire with your supplier to find out which are safe if you have pets.

Sweat Like Your Life Depends on It

Skin is a major detox organ, and when we sweat, we release significant quantities of the countless hormone-disrupting and cancer-associated chemicals mentioned above. These include flame retardants, heavy metals such as arsenic, lead, mercury, aluminum, and cadmium, and plastic conditioners like phthalates and bisphenol A. Of note, the quantities released via sweat are frequently higher (and sometimes *much* higher) than what we release via urine, the "usual" means of day-to-day excretion.

Exercise is obviously one way of inducing sweating, but so is using a sauna. Whether you prefer dry sauna or infrared, choose one that you can safely sit in long enough to sweat, and remember to drink plenty of water before, during, and after, since sweating also causes you to shed important trace minerals and electrolytes. Revisit page 90 for a primer on the other benefits of sauna use.

REPLENISHING ELECTROLYTES AFTER SWEATING

You aren't just bidding adieu to heavy metals and other toxins when you sweat; you're also losing small amounts of certain minerals that you need for good health (calcium and magnesium, for instance). By far those that we lose the most of when we sweat are sodium and chloride. Where do we find these important compounds? Regular old table salt, which combines both. In one liter of sweat, you can lose anywhere between 460 to 1840 milligrams of sodium, which is the equivalent of ¼ teaspoon to ¾ teaspoon salt. After a sweaty workout or sauna session, sprinkling some salt in your water can help replenish those lost minerals, and if eating soon after, don't be afraid to salt your next meal to taste. Revisit page 25 for salt recommendations.

Eat Your Fruits and Vegetables

Fruits and vegetables are among our most powerful detoxifying foods, each containing a number of compounds that help your body both sequester and purge toxins, including heavy metals. These naturally bitter chemicals—which aren't necessarily vitamins or minerals—can prime the body's detoxification system, upregulating chemicals like glutathione (the body's master detoxifier) that can disarm and ultimately purge environmental pollutants. Keep in mind that they tend to be more abundant in organic produce.

Cruciferous vegetables in particular are a powerful tool for detoxification. When we chew kale, broccoli, cauliflower, cabbage, mustard greens, radishes, or Brussels sprouts (all examples of cruciferous vegetables), our teeth break apart the plant's cell walls. This causes a chemical called glucoraphanin to combine with another, called myrosinase. The enzymatic marriage gives birth to a new compound: sulforaphane. Toxic to insects, sulforaphane elicits a defense response in humans, activating enzymes that help you to neutralize and excrete environmental toxins.

Cooking inactivates myrosinase, which is required to create sulforaphane. Certain gut bacteria are able to convert some of the remaining glucoraphanin to sulforaphane, but you can take matters into your own hands with a simple and powerful hack: add 1 gram of mustard seed powder (about half a teaspoon) to your veggies *after* cooking. Mustard itself is a cruciferous vegetable, and it therefore contains myrosinase. By adding the powder, you regain the ability to produce sulforaphane (plus, it's tasty).[57] Try to incorporate both raw and cooked crucifers. Or give young broccoli sprouts a try: they contain up to one hundred times the sulforaphane-producing capacity of adult broccoli![58]

Cruciferous vegetables also contain other compounds that are directly involved in detoxification, including cyanohydroxybutane and diindolyl-methane, or DIM for short. Both of these compounds are involved in your body's own detoxification processes—no expensive juice kits or cleanses required.

Load Up on Nutrient Density

Choosing a varied array of nutrient-dense foods, including fruits and vegetables, will not only help avert deficiencies, it will give your body the tools it needs for a counterstrike against various toxins. Choose foods like dark leafy greens and other fibrous veggies (kale, spinach, and arugula, for example) that are high in antioxidants as well as essential minerals, which may reduce uptake of heavy metals from the digestive tract. In one 2014 review of nutritional strategies to combat toxic exposure, researchers concluded: "Evidence now shows that a person's nutritional status can play a key role in determining the severity of environmental toxicant-induced pathologies such as diabetes and cardiovascular diseases."[59] Revisit chapter 1 for a dietary plan that focuses on nutrient density.

Eat the Smelly Foods

Cysteine is an amino acid that provides sulfur, a smelly molecule that happens to be crucial for your body's detox pathways. Foods rich in this compound include all high-protein foods—beef, fish, poultry, eggs—and vegetables such as broccoli, garlic, Brussels sprouts, cauliflower, kale, watercress, and mustard greens. Cysteine is important as a "rate-limiting" precursor to the synthesis of glutathione, your body's master detoxifier. This means that you can only produce as much glutathione as your cysteine supplies allow. Whey protein (a popular sports supplement) also happens to be high in cysteine. One 2003 study found that whey protein increased intracellular glutathione and protected prostate cells from oxidation-induced cell death.[60]

Making some of the changes I recommend may seem daunting at first, but keep in mind that you only need to do this once. (You can also integrate these changes over time, but I always advise the "one and done" approach.) And though the initial investment into what you'll be replacing them with may seem substantial (i.e., flame retardant–free furniture), keep in mind that, like buying a bunch of spices for an exotic recipe, you'll have them for a long time. It's an investment in living a Genius Life.

Though this chapter wasn't meant to be a comprehensive guide to the countless potentially dangerous chemicals (that could fill volumes), I hope it helps you to think more critically about the modern environment. And, of course, eliminating the aforementioned toxins from your environment can provide protection for you and your loved ones as the science continues to evolve. If you suspect poisoning by any of the above compounds, be sure to consult with your doctor.

Up next, as you deepen your understanding of the mind-body connection, a swan dive into the myriad aspects of good mental health, including sleep, meditation, and even your complicated relationship with technology.

FIELD NOTES

▶ Our hormones govern everything from sexual function to brain development to the fat we carry.

▶ BPA and phthalates are plastic-related compounds that can leach out of plastic and into our food and disrupt hormone function, potentially even at very low levels (having a "nonmonotonic" dose response). Heat and acid can catalyze this leaching.

▶ House dust is a major point of exposure for nearly every toxic chemical listed in this chapter. "Wet dust" regularly or use a HEPA-filter vacuum.

▶ Ditch the flame-retardant furniture and nonstick pans.

▶ Clean up your diet and oral hygiene so that you can safely cut the fluoride from your toothpaste, which is another potential endocrine disruptor.

▶ Avoid chronic use of nonsteroidal anti-inflammatory drugs (NSAIDs), acetaminophen, acid-blocking medications, antibiotics, and anti-cholinergics. Of course, they are fine when acutely necessary.

▶ Eat fish, but ensure a high selenium-to-mercury ratio. Avoid fish with the highest levels of mercury if pregnant, lactating, or a child.

▶ By eating nutrient-dense, vitamin- and mineral-rich foods, you help your body not only detox naturally, but also avoid intestinal uptake of many of these toxins. Revisit chapter 1 for a primer.

▶ Houseplants can help clean the air in your personal breathing zone.

6

PEACE OF MIND

Globally, more than 300 million people suffer from depression, which is now the leading cause of disability worldwide. An additional 260 million are living with anxiety disorders. Many live with both. And while it is possible to be genetically predisposed to each, the majority of cases are the result of factors that are environmental.[1]

At the onset of my mother's cognitive symptoms, one psychiatrist thought that her troubles were rooted in depression. He prescribed a common antidepressant drug, which my mother took for years as her dementia symptoms worsened. Once it became clear that her disease was more severe than depression, we decided to wean her off. No one in my family had realized that once on these drugs, coming off can cause withdrawal symptoms, and so she simply stayed on them until the end.

Severe enough depression can mimic dementia. Depression-related cognitive dysfunction, also known as pseudodementia, is reversible and has little in common with Alzheimer's disease or other common neurodegenerative conditions. But seeing as how one in four women over the age of forty is now taking drugs for depression (one in ten out of the general population), clearly there is an over-prescription problem. And are they effective? For many, antidepressants work no better than a placebo, with the exception being the most severe cases of depression.

Our mental health is under attack. While only about 15 percent of the general population will experience clinical depression or anxiety disorders during their lifetime, rates seem to be on the rise.[2] At

the same time, our tools for treating depression are limited. Therefore, over the following pages I aim to put the remaining pieces of
the Genius Life puzzle in place with a focus on mental health. By
the end, you'll be equipped with the tools for a healthier brain and
happier mind.

The Master Optimizer

For most animals, life is a series of threats one after another: the
threat of not making it out of infanthood alive, of starvation, of getting eaten by a larger animal, of losing your family to the elements,
of being ill-suited to find a mate to begin with . . . the list goes
on and on. It's surprising, then, that natural selection would have
allowed one third of our time to be spent unconscious, *until* you
realize the profound value that sleep has for every single aspect of
your waking life.

Quality sleep of sufficient duration is no less important to living
a Genius Life than is eating the right foods and getting sun on your
skin. Sleep lowers your blood pressure and blood sugar, regulates
your hormones, speeds up your metabolism, and strengthens your
body. It's the ultimate antiaging tonic, and nowhere is this truer than
what it does for your brain. Sleep promotes alertness, priming your
brain to receive and store information, and sleep *loss* does the opposite; routinely getting four hours of sleep or less can add eight years
to your brain's age in terms of its cognitive performance.[3]

It is now clear that persistently poor sleep can negatively affect
every system in the body, and this occurs partly through its effect on
metabolism. Animal studies and human studies alike have shown
that reduced sleep duration coincides with reduced insulin sensitivity and worse glucose control. In English, that means you produce
more insulin (inhibiting the growth hormone discussed on page 54)
and your blood sugar is more likely to stay abnormally high when
you are underslept.

In one trial published by the Endocrine Society, all it took was one single night of sleep deprivation—from 8.5 hours of sleep to 4 hours—to cause "overnight" metabolic obesity in human subjects. The dysfunction was comparable to what would be caused by gaining twenty to thirty pounds.[4] Liver sugar and fat production increased, and blood sugar was less efficiently managed. Over time, this sugar can damage the blood vessels carrying oxygen to your brain and other organs. The good news is that these changes are often avoidable—reversible, even—and all it takes is improving your sleep.

One other way that sleep keeps your brain sharp is by giving it a bath every night, and this cleansing should not be viewed as optional. Thanks to the glymphatic system, named for the lymphatic system of ducts that it resembles, cerebrospinal fluid swooshes throughout your brain as you sleep, cleansing it of various forms of toxic debris. Think you can cheat sleep? These ducts are hidden during the day, but when you're sleeping they swell by up to 60 percent, which makes way for the cleansing fluid.

Among the debris that gets swept up are the two mischievous proteins, amyloid beta and tau, which form the plaques and tangles associated with Alzheimer's disease. When we're awake, amyloid and tau are produced in the brain—a by-product of consciousness. But sleep helps prevent these proteins from sticking around. We believe this because *lack* of sleep leads to a sharp increase in their concentrations: on one night of shortened sleep, levels of amyloid beta increase by 30 percent and tau by over 50 percent.[5] These higher concentrations, measured in cerebrospinal fluid, may increase the odds that the proteins will clump together and aggregate, forming the two hallmark features of Alzheimer's disease.

For a younger person, the prospect of dementia is abstract enough without trying to imagine invisible proteins gunking up the brain. Regardless of age, however, look no further than mental health for proof of the value that sleep holds. So many of us struggle with issues related to mental health that one in six adults now takes a medication to cope, and most do so for the long term.[6] But sleep disturbance has been linked to nearly every psychiatric condition, and a growing body of research suggests it may perpetuate depression.[7]

The sleep-depression connection may trace back to an almond-shaped region deep within the brain called the amygdala that processes negative emotions. Often called the "fear center," the amygdala helps coordinate the brain's response to uncertainty, because uncertainty always presents the possibility of risk. When we're well-slept, the amygdala is kept in check by your brain's voice of reason, the prefrontal cortex, and only during genuinely threatening events does the prefrontal cortex release its inhibitory effect. But when we're underslept, all bets are off. It's like Bruce Banner being overtaken by the Hulk and that battle is won or lost during sleep.

HOW NOT SLEEPING CAN MAKE YOU FAT

Sleep acts as a master regulator of your hormones, including leptin and ghrelin, which govern satiety and hunger, respectively. When underslept, these signals fall out of whack, causing us to eat more. Fanning the flames, sleep deprivation causes a relative disengagement of your prefrontal cortex, which manages executive functions like decision-making and impulse control—both important to keeping your cravings in check. In all, people who sleep only four to five hours a night will consume an average of 400 extra calories each day, mostly from foods that the brain finds particularly irresistible, i.e., those that combine sugar and fat. Undersleep regularly and that adds up to 150,000 extra calories per year, or forty-two pounds of extra fat—all due to poor sleep![8]

An overly sensitive amygdala interprets even minor events as major stressors, so it's not surprising that these structures tend to be more active in people with depression.[9] What may surprise you, however, is that even one night of poor sleep will make anyone's amygdala about 60 percent more reactive.[10] This explains why we get testy without enough sleep, *and* why our impulses become more difficult to control. An underslept brain is stuck on high alert, and the best remedy is simply a good night's sleep.

Getting your Z's is critical to feeling happier and becoming more stress resilient, but it may make you more charming to boot. Matthew Walker, a professor of neuroscience and psychology at the University of California, has looked at the role of sleep debt on social interaction. He found that not only does sleep debt make you less inclined to socialize, it can act as a social repellent, causing you to send cues that make *other* people less inclined to socialize with you.[11] Before turning to alcohol to lubricate social interaction, as is so frequently the case, perhaps we should look first to improving our sleep.

SCARY SLEEP STATS

Given the relationship between sleep and mental health, it's unlikely a coincidence that rates of depression, anxiety, and even suicidality seem to be increasing in tandem with our collective sleep debt. Today half of adults between twenty-five and fifty-five say they sleep fewer than seven hours on weeknights, and nearly one third say they sleep fewer than six. Some of us aren't sleeping at all: over half of millennials have been kept awake at least one night over the past month due to stress—a finding of the American Psychological Association.

How much sleep do we need? Generally, adults should try to aim for seven to eight hours per night, but try for nine to ten if you are a teen. Sleep duration is important; certain processes are favored earlier in our slumber, such as deep non-REM sleep, which is when your "brainwashing" occurs. Others occur later, such as REM sleep, which helps to fortify memory and mental health. Try to maximize your sleep opportunity nightly to ensure that you can wake up naturally. And, if an alarm clock is needed, try an app like Sleep Cycle, which can help ease the transition to wakefulness.

When it comes to sleep quality, remember that bright light sometime in the morning is critical. Ideally the light comes from the sun, but artificial light, if bright enough, can work, too. Not only will this lead to earlier production of melatonin, your health-fortifying sleep hormone, but it will allow you to enter sleep more easily at night. When it comes to your bedroom, keep it cool (about 65°F) and dark—research is beginning to show that even low levels of nighttime light (a bright alarm clock display, for example) can pass through your eyelids and disturb sleep quality along with next-day cognitive function.[12] Consider blackout curtains or even a comfortable eye mask.

Here are a few more tips that may significantly improve your sleep quality:

▶ **Embrace consistency.** Maximize your sleep opportunity by getting in bed with the intent of going to sleep at the same time every night. And stick to that time; postponing sleep can backfire, causing you to become more alert, thus promoting insomnia.

▶ **Exercise.** Regular exercise boosts sleep quality, and outdoor exercise (with simultaneous exposure to the sun) may have a synergistic effect. But even if you haven't yet made exercise a regular part of your routine, a single exercise session in the evening (at least two hours before bed) may boost the quality of your sleep.[13]

▶ **Take a warm shower or bath before bed.** The drop in body temperature once you step out should signal to your body that it's time to sleep.

▶ **Try glycine and/or magnesium.** Glycine (introduced on page 33) and magnesium may improve sleep naturally.[14] As a bonus, both are powerfully antiaging. For deep sleep, try 300 to 500 mg of magnesium glycinate (which is magnesium bound to glycine) and 3 to 4 grams of pure glycine before bed.

▶ **Use your bed for sleeping and sex only.** As soon as you wake up, get out of bed, and don't return to it until you want to go to sleep for the night. No eating or burning the midnight oil from bed!

▶ **Avoid alcohol.** Even though alcohol helps you get to sleep faster, it reduces the amount of time spent in REM sleep. If drinking, sober up before bed.

▶ **Wear blue blockers for two to three hours before bed.** The light from a smartphone, laptop, or television screen can disrupt sleep, leaving you with a "light hangover" the following morning. Visit http://maxl.ug/TGLresources for a suggested pair.

▶ **Set a caffeine curfew.** Limit caffeine consumption to 4 p.m. at the latest—maybe even earlier if you are a genetically slow metabolizer (a gene testing service like 23andMe can help you determine this).

▶ **Eat more fiber and omega-3 fats.** Inflammation affects sleep quality, and omega-3 fats (found in cold-water fish like salmon, mackerel, and herring) and fiber consumption may promote deeper, more rejuvenating sleep.

▶ **Stop eating two to three hours before bed.** Ever wake up feeling crummy after a late-night meal? I have. Nocturnal eating can sabotage quality sleep.[15]

Take a Digital Hiatus

In the twenty-first century, there is no truer illustration of the double-edged sword than technology. On the one hand, it provides us with connection, convenience, and the best of the world's knowledge and entertainment at our fingertips. On the other, it is quickly becoming a replacement for real world interaction. It's no wonder technology addiction—like food and porn addiction—is on the rise.

Smartphones are a prime example of technology's good and dark sides. Similar to hyper-processed foods, which can hijack your hunger, smartphone apps are designed by developers to have the same effect on your attention. These apps are addicting in part because they lead to a stimulation of the neurotransmitter dopamine, which is involved in feelings of reward. Though posting on Instagram might not seem like doing drugs, that's essentially what a social media addict is doing from the standpoint of his or her brain.

The consequences of smartphone addiction are real. A growing body of research is finding that important aspects of our lives, including sleep, self-esteem, and our relationships, are now being compromised due to our addiction to smartphones. It's also affecting our brainpower, *even* when we are not actively using our phones. Research out of the University of Texas at Austin found that the mere presence of a smartphone during a cognitive task undermines thinking skills like memory and problem solving.[16] The investigators found what we all intuitively know to be true: that smartphones, when nearby, "exert a gravitational pull on the orientation of attention."

One other consequence of smartphones is they create a lot of unnecessary stress. Not checking our phones when we are addicted to them causes a surge of cortisol, the energy-releasing hormone that gets us up in the morning but also rises during stress.[17] Chronically elevated cortisol, a consequence of persistent stress, is not good for us; it suppresses the immune system and tears down the body's tissues in a land grab for energy. The laundry list of health conditions associated with chronically high cortisol include obesity, type 2 diabetes, heart disease, and dementia.

In the short term, abusing drugs may relieve stress for their users, but that isn't the case for smartphone addiction; checking them creates even *more* stress. Unrelenting notifications provide a recipe for mental anguish, and that's before we refresh our social media feeds. Our hyper-curated profiles now allow us to share only the best moments of our lives, filtered and edited like a highlight reel. For all but the most resilient of us, this can cause anxiety and depression once we begin comparing ourselves to others. We experience "FOMO"—fear of missing out—on a regular basis and judge ourselves as a result.

One trial, published in the *Journal of Social and Clinical Psychology*, sought to establish whether social media is actually making us depressed. The scientists formed two groups of college students. One was allowed to use their smartphones as they normally would, while the second group had their social media time cut down to thirty minutes a day. After three weeks, the results were clear: cutting down on social media led to fewer depressive symptoms, *especially* for those who were most depressed at the study start.[18] Loneliness also decreased with less social media, proving that the Social Network is no replacement for a genuine social network. The researchers summed it up: "Limiting social media use to approximately thirty minutes per day may lead to significant improvement in well-being."

Social media and mobile technology aren't going anywhere, so perhaps the key to healthily utilizing each is balance. Here are some tricks that you can use to cut down on your social media time:

▶ **Set a limit to your social media usage.** Try thirty minutes to an hour every day. Most smartphones have time tracking apps that you can use to check in with yourself.

▶ **Find activities to do without your phone.** When you go to the gym, for example, leave your phone in your locker.

▶ **Turn off notifications.** App notifications have become a scourge for mental health. Turn off notifications for all nonessential apps—*especially* social media apps.

▶ **Take a social media "shabbat."** Pick a day of the weekend to put your phone away. If you have a significant other, you might even hide each other's phones for a predetermined time period.

▶ **Unfollow, unfollow, unfollow.** Do an inventory and unfollow accounts that make you feel inadequate. If it's in your feed, it's in your head.

▶ **Fine-tune your algorithm.** Social media algorithms tend to allow extreme viewpoints to surface and promote negativity more than positivity (people are more inclined to comment on a negative or erroneous post, giving it preference by the algorithm). This can create stress.

Find Your Noble Aim

Today, many people feel burnout, which is a cluster of symptoms that includes exhaustion, alienation from work, and impaired performance. According to a 2012 study, almost one in three working adults has had such symptoms, although rates vary by profession. (Forty percent of our hard-working physicians, for example, experience burnout.) This can lead to alcohol abuse, relationship problems, and suicidal ideation.

One tried-and-true way to guard yourself against burnout is to seek a vocation rather than a job. The second is what you do for money, while the first is what you do for yourself. A vocation is work that combines both your purpose and your pleasure. It should tie together that which you naturally excel at, are passionate about, and enjoy doing, and fulfill a societal need. This "need" doesn't have to be charitable; it could merely be a problem you wish to solve. And if you have a job that you don't love but it pays the bills, that's okay, too. Try to see where it helps others (even your own family), which can give more purpose to how you spend your hours on the clock.

Finding the meaning in your work is what psychologist Jordan Peterson calls a "noble aim." This is a grand vision that ensures that the ups and downs of your journey (inherent in any career) become more tolerable, because you've put your ego in the backseat. Melissa Schilling, a professor at NYU and a noted authority on innovation, revealed a similar concept to me. In her analysis of the commonalities among breakthrough innovators such as Marie Curie, Albert Einstein, and Thomas Edison, each had an "idealized goal"—something larger than themselves that they dedicated their lives to.

Personally, when I saw what little help medicine was able to offer my mom, I wanted to know why so that I could help others who could end up in her shoes. I began talking to anyone who would listen about my experience and research. I knew with unmistakable certainty that people needed help, and they weren't getting it through the traditional channels. Helping people avoid chronic illness (and feel great in the process) became my noble aim.

Pursuing your noble aim won't be easy. In my case, every "yes"—whether it meant the appearance on a new podcast or TV show—served as validation that I was on the right path. At the beginning, however, there were a dozen *no*s for every *yes*, but I saw those nos not as closed doors but as mere speed bumps. I knew that I had something to offer, and those who'd stand in my way just didn't have the faith that I did that better health was possible.

What might your noble aim, your idealized goal, be? Maybe it's looking to change health education in impoverished communities, or to revolutionize how people travel. Or perhaps it's seeing your art reach (and change the lives of) a wider audience. Maybe it's being able to offer higher-quality food to your children. Having such goals is an *intrinsic* motivator—it comes from within—and offers far more powerful motivation than extrinsic forces, such as earning money. When you are intrinsically motivated, the money often follows.

One other benefit of having a noble aim is that it keeps you focused on the journey, not the outcome. Attainment of any goal—whether it is reaching a higher salary or seeing the car of your dreams in your driveway—inevitably leads to hedonic adaptation. In other words, you get used to it. This keeps you constantly wanting, which Buddhists know is the source of all suffering. Instead, allow your life to be a practice, similar to yoga, meditation, fitness, or playing a musical instrument: chase *progress*, not perfection.

Seek Novelty

Under normal circumstances, we can predict how we might feel standing in line at our usual coffee shop or stuck in traffic on our regular route to the gym or office. This is because familiar settings inspire familiar thoughts. Throw yourself into a novel environment—even if it's just a different coffee shop than your usual haunt—and that predictive ability goes out the window. Novel settings inspire novel thoughts, and there's no better way to experience a novel environment than by traveling.

The health benefits of travel are numerous, though difficult to quantify. We know that in the brains of mice exposed to novel environments, there is a fourfold increase in the creation of new connections between brain cells compared to the brains of mice allowed to stagnate in their same old environment.[19] We also know that feelings of awe and wonder—common when exploring a new place—have been linked to lower levels of inflammation.[20] And walking or renting a bike is one of the best ways to explore a new city, making travel an incredible opportunity to be more active.

Travel also allows you to change up your diet and experience new flavors and spices, which exposes you to an array of unfamiliar nutrients. Unless you have dietary restrictions (due to allergies, for example), allow yourself to "loosen your belt" while traveling and taste the local flavors and cuisine. Chances are your indulgences will be matched by your increased activity levels—unless, of course, you're on a cruise.

Take vacations, but don't let that be an excuse not to exercise. Since odds are you will be walking more than usual (engaging in more NEAT, described in chapter 4), focus your workouts on resistance training. No gym? No problem. Lunges, push-ups, chair dips, and planks are all examples of exercises you can do right in your hotel room. This will increase the odds that any excess of energy you consume will be shuttled toward building muscle instead of fat.

BEATING JET LAG, THE GENIUS WAY

As great as travel is for the mind, it can take a physical toll on one's body. Sitting for many hours in transit is incongruous with our biological yearning to move. Crossing time zones throws off our bodies' circadian rhythms. And eating on the road presents its own set of health problems: packaged convenience foods rich in easily digestible carbohydrates, processed oils, and industrial additives can make us feel crummy all around.

When we know that our schedules are going to be suddenly thrown off by travel, we can use intermittent fasting (i.e., strategically avoiding food) to soften the blow to our circadian rhythm. I stick to water and avoid food and caffeine during overnight flights, until eating a large fast-breaking meal (along with a cup of coffee) the morning I arrive. A protocol similar to this, developed by a chronobiologist at the Argonne National Laboratory in Illinois, was shown to significantly reduce occurrence of jet lag in a 2002 study.[21]

Because your brain's natural melatonin production cycle will be thrown off, getting to sleep in your new time zone may be a challenge. Ensuring that you get bright light during the morning in your destination can help accelerate the acquiescence of your internal clock to its new time zone. Exercise is also a powerful time-setter, and performed in the morning can also help reorient your body's clock. Finally, don't be afraid to supplement with 3 to 5 mg of melatonin thirty minutes before bedtime in the new time zone, which can significantly ameliorate symptoms of jet lag.[22]

Seek Mystical Experience

Though the human mind is expansive and capable of extraordinary feats, it is also very good at maintaining order. Philosopher Aldous Huxley called the impulse for order the *reducing valve*, there to rein in conscious awareness like the blinders on a buggy horse. Our mind allows us to be organized, get stuff done, and fit neatly within the framework of society. But for some, these cognitive shackles come at a cost; excessive mental rigidity can lead to repetitive thoughts, obsession, paranoia, anxiety, and depression.

Huxley's valve might actually be a system in the brain called the default mode network, or DMN. As the seat of self-awareness, the DMN is always active and is especially so when we're idle. Higher activity within the DMN may represent a mind more prone to negative thoughts and rumination, and one more resistant to change. Treating the psyche under everyday circumstances may be like trying to hit a moving target, but taking the DMN temporarily "offline" may grant access to the root cause of many of our mental health woes.

Psychedelic drugs, including LSD and psilocybin, the psychoactive component of "magic" mushrooms, have been used for millennia to help users transcend self-awareness. Recently, researchers at Imperial College in London confirmed that these compounds also anesthetize the DMN, which may provide an opportunity to treat conditions like depression. After being given the drug psilocybin, subjects' self-awareness dissipated into a feeling of oneness with the world around them—a sensation termed *ego-dissolution*—and brain scans revealed a simultaneous decrease in activity in the DMN.[23]

At Johns Hopkins University School of Medicine, researchers wanted to see whether the same drugs, by reducing activity within the DMN, could help alleviate depression in patients with life-threatening cancer. Many cancer patients experience depression; nearly half develop major depression or other mood disorders. After subjects received the hallucinogen, they were instructed to lie down on a couch while they were played soothing music through a set of headphones. Each were carefully monitored and asked to close their eyes and "trust, let go, and be open." The results of the experiment were staggering.

Afterward, 80 percent of patients experienced significant reductions of anxiety and depression. Most encouraging, the results were not fleeting; the same proportion of patients experienced continued relief from depression six months *after* treatment. When describing their experience, many subjects called it "spiritual." In fact, patients with the *greatest* success reported having experiences that were among the most meaningful of their lives—up there with having a baby. Prior to the study, some of the patients had become "myopically focused" on their cancer diagnoses, but afterward, many reported that their fear of death had been erased.[24]

More research is needed to elucidate which patients will benefit the most from this treatment, as well as to develop protocols that minimize risks. Still, the disengagement of the DMN and subsequent mystical experience seemed to do the heavy lifting in helping the patients beat their depression. So while the drugs may have been a powerful catalyst (and indeed necessary for some), maybe regular ego-dissolving experiences such as meditation can help us achieve some of the benefits of psychedelic therapy, without the psychedelics.

Here are some ways that you might safely quiet your default mode network:

▶ **Meditate.** Meditation may calm an overactive default mode network, as studies show that regular meditators have reduced activity in the DMN. More on meditation shortly, along with a simple practice to get you started.

▶ **Try breathwork.** Breathwork—consciously manipulating your breath under controlled conditions—has been a part of certain yogic traditions for millennia, and can bring about an altered state of consciousness. Numerous studies have suggested a benefit for depression, anxiety, and trauma.[25]

▶ **Shut off your senses.** Sensory deprivation tanks have become a popular pastime for their ability to alter consciousness. The deprivation of external sensory input often evokes visual or auditory hallucinations.

Mind Your Sensory Input

Living in New York City, I became exquisitely aware of just how stressful ambient noise can be. It often shocked me the volume of noise one must endure just walking on the sidewalk or taking a subway. I wondered about the damage I was doing to my hearing by being a bystander, but it also dawned on me that the noise likely affects more than just one's ears. Upon deeper inquiry, I discovered there are major health consequences to what has become an increasingly common problem: noise pollution.

For much of our ancient past, loud sounds usually forewarned of danger. A lion's roar, a snake's hiss, the cacophony of a waterfall, or the scream of a loved one (or oncoming enemy) would each require our immediate attention. Sudden loud sounds are far more jarring than objects that pop into our vision, and it's only when both are combined that we jump from our seats during horror movies and thrillers. But off screen, daily life has become excessively noisy, and modern life can elicit the same stress response as a predator would, often without our consent.

Unlike our other senses, we can't consciously disengage our hearing, and this is unfortunate, since experiments in humans have shown that loud noise can cause a release of various stress-related hormones, including cortisol. This may be why chronic noise exposure close to one's home coincides with higher rates of heart disease and type 2 diabetes. One Danish study found that for every ten-decibel increase in traffic noise in proximity to one's residence, the risk of developing type 2 diabetes increased by 14 percent.[26] While residential areas with lots of traffic noise may also be heavily polluted or areas of low income, this study controlled for those variables and found the relationship held up nonetheless.

Perhaps the most noticeable cost of noise pollution is our mental health. Intrusive noises are annoying, and research suggests that noise annoyance can take a serious mental toll. In a study published in the journal *PLOS ONE*, frequent exposure to aircraft noise was associated with a twofold higher prevalence of depression and anxiety in the general population.[27] Depression is the single largest contributor to disability worldwide, and the World Health Organization (WHO) estimates that Western Europeans alone lose up to 1.6 million years of healthy living annually from noise pollution.

Stress hormones also affect learning and creativity (more on this in the following section), and our children may be particularly vulnerable to the effects of loud noise exposure. The WHO reports that children living in areas with high aircraft noise have not only higher stress levels but delayed reading ages and poor attention. As more and more children are put on stimulant medications for their perceived learning deficits, perhaps a look at their environment is warranted before moving on to more intensive treatments.

How might we cut down on our exposure to excessive ambient noise? Moving to a quieter neighborhood (or higher floor) may be among the most effective, but also least practical. Here are some other tips you can use to effectively blot out excessive, stress-inducing noise exposure:

▶ **Invest in noise-canceling headphones.** These are a lifesaver in noisy environments, like on flights. Just avoid wearing them while driving or in an urban environment where they could impair awareness of your surroundings.

▶ **Use earplugs.** Earplugs can be effective for blocking out sound during sleep, when using loud appliances, or during meditation.

▶ **Sleep with a pillow on top of your head.** This works wonders (at least for me) if you are a side or stomach sleeper.

▶ **Use a white noise machine or fan.** Humans are able to get used to persistent noise. If the tone is consistent and kept at a low-enough volume, white noise can be used to mask annoying environmental sounds. Air conditioners and fans work well for this purpose.

▶ **Avoid loud restaurants and bars.** I suspect that one reason we feel called to imbibe in bars and restaurants is that—beyond social pressure—they're simply stressful to be in due to their noisiness. Seek quieter establishments.

Might noise have played a role in my mother's ill health? I'll never know, of course, but one thing is certain: she spent the vast majority of her life in New York City's midtown, where noise pollution is profound.

Meditate, Don't Medicate

Before I began a meditation practice, I thought what you're probably thinking right now: "I can't clear my head for a few seconds, let alone minutes every day!" But meditation isn't about clearing your head; it's about making friends with your mind. Similar to swimming, or riding a bicycle, meditation is a valuable skill to have whether it becomes a regular part of your day or not.

Humans have been practicing meditation for a long time, *long* before any bestselling books were written about it. Though its exact origins are elusive, it's believed that primitive hunter-gatherer societies discovered it while staring into the flames of campfires. Over thousands of years, meditation evolved into a structured practice, with Indian scriptures called *tantras* mentioning formalized techniques as early as five thousand years ago.

Meditating has a number of benefits. Some are achieved with a regular practice, but some are won after just a single session. It can quell stress, anxiety, and symptoms of depression. It can reduce inflammation (perhaps via its positive effect on stress hormones).[28] It makes your brain work better. And it might even slow the rate at which you age by supporting your telomeres, structures that protect your DNA from damage.[29] But perhaps the most practical benefit of meditation is its effect on emotional regulation.[30]

Have you ever come up with the perfect comeback to your boss or significant other, only *after* the confrontation ended? What a shame, you say to yourself, that your witticism didn't arrive when it really mattered. This happens to the best of us, and the reason comes down to the effect that stress has on creative thinking. Long ago real stress meant physical danger—*not* interpersonal conflict—and a sharp tongue isn't going to help you outrun an angry bear. Our bodies respond similarly today, but modern stress is seldom a life-or-death ordeal.

Consider another common scenario: has a friend or family member ever irked you to the point of eliciting a harsh reaction, one you wished only after the fact that you could take back? We've all done this in the heat of the moment—said something we know we shouldn't have said, and likely didn't even mean in the first place. But because we were in the middle of a fight-or-flight response, we showed our teeth, fangs and all, and it was ugly and regrettable.

What the above situations have in common is that meditation can help prevent both of them, by widening the gap between stressful stimuli and the response it elicits. That gap is called awareness. By helping to increase awareness, meditation *decreases* reactivity, giving you the space to be your best, most creative and compassionate self during those inevitable times of psychological stress.

FAQ

Q: Do meditation apps work?

A: Meditation apps can be helpful, but I'm not a big fan of relying on them because doing so keeps you tethered to your mobile device, and I believe we all outsource too much to our devices already. Learning to meditate (without using your device as a crutch) gives you the power from within; it's akin to learning how to fish as opposed to being given a fish—and we all know how that goes. Thankfully, there are now many meditation teachers, so find one in your neighborhood who can coach you. Better yet, sign up for an in-person or online course—visit http://maxl.ug/TGLresources for suggestions regarding the latter.

To begin with a simple meditation, turn off any music and silence your phone. Noise during meditation is not a deal-breaker, but you'll want to avoid unnecessary distraction as best as you can. Find a place to sit comfortably. You can have your back supported. Take a deep breath and settle in. Make sure your arms are uncrossed. With your eyes closed, focus on your breath, the gentle inhale and exhale.

As you draw your awareness to your breath, your thoughts are going to want to wander—to your shopping list, to your kids' whereabouts, to the unread emails and text messages waiting in your inbox. This is *totally okay and normal*. All you do when this happens is—*without scolding yourself*—direct your attention back to your breath. Thoughts are normal. As my friend Emily Fletcher, a globally renowned meditation teacher, always says, trying to will your mind to stop thinking is akin to trying to stop your heart from beating.

Because thoughts will naturally arise, the objective of meditation is not to clear your thoughts. It is, in fact, to *allow* your thoughts to arise, perhaps even to acknowledge them, and then to gently and gracefully place your awareness back onto your breath (some practices use a mantra instead of breath—the idea is the same). Imagine saying, "Hey, thoughts, I love you, but I'll see ya later!" This action will relax you and strengthen your mind, allowing you greater space between awareness and reactivity, which is so useful in everyday life.

Try meditating for ten to fifteen minutes a day, and make an eight-week commitment to your practice. One trial from New York University found that this exact protocol decreased negative mood and anxiety and enhanced attention, working memory, recognition, and emotional regulation in nonexperienced meditators compared to controls.[31] In terms of time of day, most people will find first thing in the morning (prior to consuming any coffee or food) to be the best time.

Remember: there's no such thing as good meditators or bad meditators—it's a practice, and the more you do it, the more you will reap from it.

Avoid Wobbly Furniture

Someone once said, "It's easier to act your way into a new way of thinking than to think your way into a new way of acting," alluding to the ability of our actions to shape our thoughts. This idea has received scientific support, with studies clearly showing that the mere act of smiling can invoke feelings of happiness *and* reduce the physiological effects of stress. Even when our minds are working against us, it seems, we can still will them into a more manageable state.

Our spaces also shape our thoughts. In chapter 3, you discovered the benefits of nature exposure on our mental and physical health.

Nowhere are the consequences of our lack of nature exposure more visible than in our cities. According to the National Recreation and Park Association, physician-diagnosed depression was 33 percent higher in the residential areas with the fewest green spaces, compared to the neighborhoods with the most. Physician-diagnosed anxiety rates were 44 percent higher.

City or country, your immediate surroundings—including your furniture—possess domain over your thoughts as well. Practitioners of the ancient Chinese art of feng shui (translated means "wind" and "water") have known for millennia that the arrangement and placement of objects in a space can affect one's health and well-being. Proper placement, optimizing the flow of space as if it were water, can promote feelings of inner harmony, while poor placement can cause a disruption of that sense of balance. And just as it has with meditation, science is slowly catching up with the wisdom of the ages.

In one groundbreaking study performed at the University of Waterloo in Canada, researchers sought to find out how furniture may alter one's emotional life. They placed individuals at one of two tables: one where the furniture was wobbly and one with stable chairs and table, and then showed subjects photos of celebrity couples. Those who sat at the wobbly table projected that couples were more likely to break up than those seated at stable furniture.[32] The furniture even altered subjects' value systems: when asked to describe what they admired in a relationship, those who sat at wobbly furniture prioritized stability-promoting traits, like reliability, whereas those sitting at wobbly furniture favored instability-promoting traits, like spontaneity.

With 93 percent of our time now spent indoors, curating our environments may be more important than ever.

Final Principles

As important as good health is to a Genius Life, how you live your life is just as critical. This is a fact my mother instilled in me, and in the following section I hope to instill in you the principles she left behind.

Be Kind to Animals

My mother was a lifelong advocate for animals. She made the decision to not eat meat except on rare occasion, and even then it was only lean poultry or fish. Witnessing her sacrifice made me understand that our first moral responsibility should be to ourselves and our loved ones, and this should include nourishing ourselves with the nutrition offered by animals.

Can you eat meat and still stand for the rights of animals to be treated with dignity and respect? Absolutely, and you should. Whenever you can, vote with your wallet for a farming system that treats animals and the environment more kindly; look for "pasture-raised," "100% grass-fed," or "humanely raised" on packages (note that "organic" can merely mean organic grain fed to feedlot animals and is therefore not sufficient to suggest humane treatment). Avoid foods that promote unnecessary suffering for animals, such as foie gras or formula- or milk-fed veal.

When it comes to cosmetics and home cleaning products, always support cruelty-free brands that do not test on animals—usually this will be indicated on the product label. You don't need a face lotion that had to be squirted into the eyes of rabbits to be determined "safe"; stick to the kinder options out there, as there are plenty. (You've already learned about other safety checks in chapter 5.)

Finally, give a pet a home; there are countless homeless dogs and cats in need of a place to live. Never buy an animal; not only are there perfectly good and deserving animals in your local shelters, but buying puppies often supports puppy mills. These are inhumane operations for many reasons. Plus, you're more likely to adopt mutts, which are more interesting and tend to have fewer of the health problems associated with purebreds.

Be Kind to People

Kindness begets kindness. Life is a whirlwind, and it leaves many traumatized. Even the "luckiest" among us get to see their loved ones age, get sick, and pass away. Be empathetic to the suffering in others and try to ease their burden wherever you can. If someone is unkind to you, realize that it probably comes from a place of deep suffering, perhaps even deeper than you can imagine. Respond with kindness.

Befriend the Lonely

Loneliness is pervasive today. A 2018 survey by the health insurer Cigna found that nearly half of all people in the United States feel lonely, with younger people bearing the brunt of the pain.[33] It doesn't just make us sad, either. Recent research presented at a conference of the American Psychological Foundation found that people with weak social ties had a 50 percent higher risk of dying early.[34] My mother always encouraged me to befriend the lonely. Look for the awkward or excluded and say hi. You'll make a new friend and perhaps save a life in the process.

Stand Up Against Bullies

Strength is a virtue, but not everyone is in a position to be strong—and not even the strong are *always* strong. If you see someone abusing their power, bullying, or taking advantage of someone else, stand up for that person. If you're a parent, raise your kids to do the same. Childhood bullying is on the rise: more than one out of every five students report being bullied, according to the National Center for Education Statistics.

Give to Those with Less Than You

I was once sitting in a New York City subway car when I heard the door in between cars open up. Without looking, I thought callously, *another panhandler.* When I turned to look, it was a man with such severe burns he had no nose or ears. His hands, which he used to hold on to the rails of the moving car, had no fingers. He had a tin can around his neck and was asking for charity.

In the moment, I was awestruck by the man's terrible condition. I had only twenty-dollar bills in my wallet, and I had never given such a high denomination to a beggar. I felt paralyzed, and it was easy for me just to give nothing, so that's what I did. He passed, moving on to the next car. I failed to act charitably, and I've regretted it ever since.

Be charitable. Whether that means giving away money, time, or things, someone less fortunate could always use what you have. Every fall or spring, go through your dresser and closet for clothing that is no longer serving you and donate it to a local nonprofit. Do the same in your kitchen.

Never Lie

In his classic book *The Four Agreements*, don Miguel Ruiz urges readers to be impeccable with their word. Dishonesty of any sort, according to his philosophy, creates suffering and limitation in one's life. I don't believe my mother ever read Ruiz's work, but she nonetheless raised my brothers and me to always tell the truth.

Always make an effort to tell the truth, even if it hurts. And be truthful with yourself. Self-deception is a tool we often use to cover up our problems, but it will hurt you in the long run. As children's book author Dr. Seuss once wrote, "Be who you are and say what you feel, because those who mind don't matter and those who matter don't mind." Always be truthful.

If You Can Teach, Teach

We all acquire different skills that could be of value to others. Too frequently we sequester these abilities, employing a scarcity mind-set as if sharing our gifts somehow lessens their potency or value. Enjoy opportunities to teach what you know to others. This has the added benefit of deepening your own fluency; as is often said, the best way to learn is to teach!

Never Take Advantage of Anyone
(or Let Anyone Take Advantage of You)

My family always flung around Yiddish terms—that's what growing up Jewish in New York will do to you. And though many of the terms were lost on me, I always remembered "Don't be a schnorrer." Essentially, it means *Don't be a sponge*. Generosity is wonderful, and when you are on the receiving end of it, don't take advantage of it. Similarly, be generous but don't let anyone take advantage of you. Set clear boundaries. Respect yourself enough to know when to say no.

Admit When You're Wrong

It's okay to be wrong; no one is right 100 percent of the time. But few traits are worse than stubbornness when you're wrong. If you've made a mistake, either professionally or personally, cop to it.

Be Brave

Bravery was a virtue in my home. It can take many forms: going in for a medical test or procedure that you have been avoiding, giving that presentation, or sharing an uncomfortable truth with one who needs to hear it. My mother always applauded bravery in all of its many forms. Be brave and encourage and commend it when you see it in others.

Express Gratitude

My mother never felt sorry for herself in all of the eight years that she battled with her strange manifestation of dementia. And if she did, she never expressed it to me. In fact, she told me once, "I'm happy that I've gotten so far, so far." Being grateful for what you have doesn't mean you can't aspire to greater. But no matter where you are relative to where you want to be, express thanks for what you have. Many have less than that.

Be Humble

I know you're awesome. But stay grounded. We don't live in a vacuum, and we each rely on one another for various needs at different times in our lives. Plus, we each have different skills and talents. You may do the best "Purple Rain" in the karaoke bar, but the tone-deaf singer next to you may know how to do a tax return like nobody's business. Who's to say that one is better or more virtuous than another?

Change Your Thoughts

Getting the results you want—whether it's a stronger body, a sharper brain, or a healthier relationship with food—usually requires a change in behavior, but most people jump into action without changing the core beliefs which have led to the unwanted behaviors in the first place. It's a folly illustrated by New Year's resolutions—most people vow to change their behavior (joining a gym or starting that new diet) but tend only to have short-term success. For important behavioral changes to be sustained, what must precede action is a change in your core beliefs. It's the "noble aim" concept from earlier, but applied to your lifestyle. Rather than trying to eat healthier (the action), focus on the core belief that you *deserve* to feel extraordinary, and that doing so is easy and attainable. Switch your beliefs, check your language, be mindful of the present moment, and results will follow.

Just as there is no one-size-fits-all diet, there's no one-size-fits-all prescription for the optimal lifestyle. We each bring to our schedules social, professional, and personal obligations. But with the above guiding principles, many of our most pressing mental health challenges should come into reach.

In the following chapter, I will put all of the pieces of the *Genius Life* together for an overarching plan, plus a shopping guide to load up on foods to support your best life yet.

FIELD NOTES

- ▶ Sleep is a powerful regulator of many aspects of your health. It can influence your mood, weight, and even your social life. Don't skimp on sleep!
- ▶ Take regular vacations from technology, which studies show can perpetuate depression, particularly once we become addicted.
- ▶ Aim to find your higher calling, or noble aim, for happiness in your professional life. This allows you to remove your ego, and instead pursue your North Star.
- ▶ Travel often. It boosts your health and may breathe life into your romantic and intellectual life.
- ▶ Seek experiences that put your ego in the back seat, which may have a rewarding effect on your mental health.
- ▶ Be cautious about excessive noise exposure. You can't disengage your sense of hearing, making you susceptible to noise-induced stress.
- ▶ Meditate regularly, or at least know *how* to meditate so that you can dose yourself accordingly.
- ▶ Be mindful of the ability of your environment to shape your thoughts. What may seem like anxiety or depression could just be a mental reflection of a nerve-racking environment.

7

PUTTING IT ALL TOGETHER

We've delved into the science of healthy eating, circadian biology, exposure to nature, exercise, toxic everyday chemicals, and sleep—an all-encompassing guide to living the Genius Life. Some of the things you will immediately notice once you begin integrating these findings are increased energy, less brain fog, and a happier and less anxious mood.

But integrating all of these concepts into a twenty-four-hour period can be intimidating for the uninitiated. Over the following pages, I'm going to show you how to work each learned principle into your day. You'll get an updated shopping list so that you may stock your kitchen with delicious and nourishing whole foods, and I'll provide a framework for you to determine your "Custom Carb Score" so that you can assess your tolerance for higher starch foods. Then I'll follow with an example "24 Hours of Genius."

The first step on the path to living a Genius Life is reengineering your environment to support your success. I call this "Laying the Foundation."

Week 1: Laying the Foundation

In week 1, you'll be curating your immediate environment to support healthy stress and hormone regulation, which will help you feel better, while also promoting the mental fortitude you'll need to make dietary shifts in week 2. These include clearing out endocrine-disrupting chemicals, optimizing your sleep, and shedding stress.

Clearing Out Endocrine-Disrupting Chemicals

In chapter 5, you discovered the pernicious nature of many of the chemicals that we interact with every single day. These include plasticizing chemicals like phthalates and bisphenol A, along with flame retardants and parabens. Remember that these chemicals can interfere with hormones, which guide *everything*. We are particularly vulnerable to their disruption when we're young and still developing, but hormone disruption in later life can come with potential consequences, too; think elevated hunger levels, a predilection for fat storage, metabolic disease and cancer, low libido, and reduced fertility to name a few.

While eliminating all potential sources of exposure is impossible, the following framework should help you minimize your day-to-day exposure. Begin by discarding all of the below (I'll tell you what to replace them with next):

▶ **Plastic food/beverage containers.** Plastic Tupperware, plastic bottles, plastic water-boiling devices, coffee makers with plastic filter carriages, plastic food wrap, any plastic that comes in direct contact with heated foods or beverages.

▶ **Flame retardants.** Flame-proof furniture, clothing with flame retardants (children's clothing usually will come with a warning that they are not flame resistant).

▶ **Cookware.** Nonstick coated cookware, especially if old and worn.

▶ **Bathroom supplies.** Dental tape, cosmetics with "-paraben" chemicals, chemical sunblocks.

▶ **Synthetically fragranced products.** Anything with "fragrance" that is not explicitly plant-derived. This includes cosmetics, laundry detergents, cleaning solutions, and fabric softeners.

▶ **Receipts.** Avoid handling store receipts. These are usually covered with bisphenol A. If needed, wash hands thoroughly after handling.

Aside from getting rid of the above items, make it a part of your usual routine to vacuum and wet dust regularly (this is simply dusting with a damp cloth, as opposed to a dry feather brush, which simply redistributes hazardous, chemical-laden dust).

Stock Up on Endocrine-Safe Products

By replacing the above-mentioned items with the ones below, you can play it safe where your finely tuned system of hormones is concerned. And once you make the initial investment, the return on your health will be priceless.

▶ **Glass food and beverage containers.** You can buy an inexpensive Pyrex (hardened glass) set, which will last years. If you are concerned about breakage, opt for stainless-steel storage. Don't worry about plastic lids if they don't come in contact with warm foods or beverages.

▶ **Flame retardant–free furniture and clothing.** As long as you have a smoke detector in your home, you are just as safe from fires.

▶ **Use a reverse-osmosis water purifier.** If you are concerned about your water quality, using an RO purifier can greatly cut down on exposure to contaminants.

▶ **Use naturally fragranced products.** You can continue to enjoy nice fragrances; just make sure they come from essential (derived from the word *essence*) oils from plants.

▶ **Use a reusable glass or stainless-steel water bottle when on the go.** Many public places, including airports, now offer filtered water bottle refilling stations.

▶ **Keep plants in your home.** Just make sure they're pet safe if you have pets. Revisit page 167 for some suggestions.

Optimize Your Sleep

In *Game of Thrones*, it isn't until the Night King is slayed that all of the zombie White Walkers get defeated. Improving your sleep may as well be considered killing the Night King, because it's the one thing that helps you to optimize all of the other areas of your life. You can revisit page 47 for a thorough guide, but the keys include:

- ▶ **Practicing sleep hygiene.** Keep your bedroom your sleep sanctuary; ensure that it is cool, dark, and free of jarring noises.
- ▶ **Avoiding bright blue light at night.** Blue-blocking glasses can help, or just simply turn off your devices and dim the lights around your home in the hours before bed.
- ▶ **Aim for a consistent sleep schedule.** If you stay up two to three hours past your normal bedtime on Friday and Saturday nights, that's the equivalent of crossing time zones every weekend. And you wonder why Mondays are often so difficult; you're essentially starting the week, *every* week, jet-lagged!
- ▶ **Optimize your day for sleep.** Bright morning light, exercise, and non-exercise activity all facilitate healthy sleep the following night.

Practice Stress Hygiene

Being able to both avoid and diffuse your stress is crucial to living a Genius Life. Chronic stress—different from the type of temporary stress you impose on yourself during a workout—can impair your immune system and your memory, while also making you less creative. It can also negatively affect your waistline. Stress causes many people to overeat, but it also impairs nutrient absorption. And eating while stressed can negatively affect digestion, causing unpleasant symptoms like bloating and diarrhea (many people experience the latter when giving public speeches for this reason).

Here are some effective ways of practicing stress hygiene:

- ▶ **Exercise.** Exercise is medicine! It lowers stress hormones and increases chemicals called endorphins (also released during sauna sessions) that improve your mood and act as natural painkillers.

- **Take control and say no to things.** Check in with your true desires and needs and don't be afraid to say no to asks of your time that impose on your health and well-being. Don't stretch yourself thin or take on more than you can handle. Prioritize "me time."

- **Limit screen time.** Technology can be stressful. Set moratoriums on your screen time and partake in activities that require you to put your device away. Some examples include bicycle riding or leaving your phone in your locker during a gym session.

- **Reduce news consumption.** The "news" today is intentionally alarmist because it commands your attention, which can then be manipulated for ad sales. Reduce your consumption of news both on TV and on social media.

- **Set the mood.** Who said you can't set the mood for *yourself*? Tidy up your house as if you have a date coming over, but do it for yourself. Use aromatherapy from essential oils (*not* synthetic drugstore fragrances), which can reduce anxiety,[1] and craft a playlist of calming but uplifting music.

- **Develop a meditation practice.** Revisit page 192 for a simple meditation. Meditation can be used regularly or "dosed" on an as-needed basis.

- **Breathe.** Simply taking a few long, slow breaths can de-excite your nervous system, bringing you into a "rest and digest" phase. This is useful before meals or any time you want to draw your awareness to the present moment.

Weeks 2 to 3: Focus on Food

Clear Out Your Kitchen

There are good reasons to minimize consumption of grains and grain products made with wheat, corn, and rice. These foods, which now make up over half of the calories we ingest every day, are nutrient-poor and calorie-dense; this means that they help fuel not only the obesity crisis but our widespread nutrient deficiencies. What does this mean for you? Greater difficulty in dropping the weight or keeping it off, and lower defenses against stress and aging.

Many grains today are processed into grain *products*, which comprise the widest category of easily accessible packaged and processed, *hyperpalatable* foods. These foods are a slippery slope, and are difficult to consume in moderation. They often come with tagalong endocrine-disrupting chemicals (leached from packaging), residues of glyphosate (which is sometimes used as a desiccant in grain farming), and heavy metals. And they're poor choices from a dental health perspective. While you may choose to enjoy the occasional whole grain side dish, the optimal amount of *refined* grain products is zero. Here is a list of grain products to avoid:

▸ **Processed grain products:** Breads, pastas, wraps, cereals, baked goods, noodles, soy sauce, chips, crackers, cookies, oatmeal (exception: gluten-free, steel-cut), pastries, muffins, pizza dough, doughnuts, granola bars, cakes, pancake flours and mixes, juices, fried foods, and frozen packaged foods. Anything with rice flour, wheat flour, enriched wheat flour, whole-wheat flour, or multigrain flour in its ingredients list.

BREAD: A SLIPPERY SLOPE

Bread may be one of humanity's oldest and most revered processed foods, but it is a processed food nonetheless. Most commercial breads are empty calories, save for the synthetic vitamins added in. They are also loaded with refined salt, constituting America's top source of dietary sodium, and gluten, the protein found in wheat, barley, and rye. Many wheat varieties are bred (no pun intended) to contain high levels of gluten because of its gummy, pleasing texture, and some wheat *products* (such as wraps) have even more added in. And gluten is now added to a plethora of other processed foods, including soy sauce and gravies.

Unfortunately, no human can fully digest this protein, and yet the modern world has allowed it to become abundant in our diets. This poses a number of digestive challenges for a substantial portion of the population. For instance, in people with celiac disease, a violent immune response is stimulated, including permeability of the intestinal barrier, which can let inflammatory bacterial toxins enter the blood. For those without celiac? A milder form of the same process can occur.[2]

Does this mean no bread *ever again*? If you do not have celiac disease or a wheat allergy, you may be able to include some in moderation, though I suggest avoiding it completely if you have any type of autoimmune or inflammatory condition. In terms of "better" bread types, sprouted grains are generally best, along with sourdough, which has been fermented and contains less gluten as a result. Personally, when I'm craving bread, I'd rather opt for the occasional grain-free bread (sometimes labeled as "Paleo" bread), many of which are now on the market and are made with nutrient-dense flours from almond or coconut. Just remember that bread of any type is a slippery slope and easily overconsumed.

Snacks with added sugar and sugar-sweetened beverages make up a major source of empty calories in the Western diet. But empty calories are not benign; they encourage nutrient deficiencies, expand our ever-growing waistlines, plus all of the obesity-associated diseases. Startling research published in the journal *Circulation* estimates that nearly two hundred thousand people die each year from conditions driven by sugar-sweetened beverages, including cancer and heart disease.[3] Cutting out these junk foods (along with the refined grain products above) may be among the most impactful moves you can make for a slimmer waist.

▶ **Sugary foods and beverages:** Candy, energy bars, granola bars, instant oatmeal, ice cream and frozen yogurt, jams/jellies/preserves, gravy, ketchup, commercial salad dressings, fruit juice, fruit-on-the-bottom yogurts, sodas, commercial fruit smoothies, sports drinks, sugar- or juice-sweetened dried fruit.

▶ **All concentrated sweeteners:** Honey, maple syrup, corn syrup, agave syrup or nectar, simple syrup, both brown and white sugars.

Though the research on chemical additives is evolving, animal studies suggest that certain synthetic emulsifiers that help lend processed foods a creamy texture can degrade the gut habitat, shifting it toward a pro-inflammatory state. Inflammation in the gut doesn't stay in the gut; it affects all your other organs, including your brain.

▶ **Sources of industrial-grade emulsifiers:** Anything with polysorbate 80 or carboxymethylcellulose in the ingredients list. Common offenders include ice cream, coffee creamers, nut milks, and salad dressings.

With meat and dairy, quality rules. On the other hand, processed meats and cheeses usually contain nasty chemical additives and metabolites. One such example is sodium nitrite, a preservative used to cure deli meats. Sodium nitrite can transform into chemicals called nitrosamines, which may promote metabolic dysfunction on top of being carcinogenic (worth noting: vitamin C, which is found in fruits and vegetables, effectively suppresses nitrosamine formation).

▶ **Industrial and processed meats and cheeses:** Grain-fed red meat, feedlot chicken, processed cheeses and cheese spreads.

Avoiding common cooking oils will help your body minimize excessive consumption of omega-6 fatty acids. These fats, discussed in chapter 1, are also drivers of inflammation due to their highly delicate and damage-prone nature. They also contain man-made trans fats, which are mutant fats with no safe level of consumption. Keep in mind that these fats often hide in grain products, listed previously, along with myriad spreads, dressings, and cooking sprays.

▸ **Commercial cooking oils:** Margarine; buttery spreads; cooking sprays; and canola, soybean (sometimes labeled "vegetable oil"), cottonseed, safflower, grapeseed, rice bran, wheat germ, and corn oils. These oils are often included in various sauces, mayonnaise, and salad dressings. (Even if they're organic, toss them.)

▸ **Nonorganic, nonfermented soy products:** Tofu.

▸ **Synthetic sweeteners:** Aspartame, saccharin, sucralose, acesulfame-K (also known as acesulfame potassium).

A NOTE ABOUT OILS

As I mentioned in chapter 1, the staple oil in your kitchen should be extra-virgin olive oil, which combines a healthy arrangement of heart-healthy fats and powerful, health-promotive phytochemicals. In fact, a compound in extra-virgin olive oil has been demonstrated to possess anti-inflammatory potential on par with low dose ibuprofen, without any potential for negative side effects. Go organic if you can afford it, as organic contains 30 percent higher levels of this anti-inflammatory chemical compared to conventional extra-virgin olive oil.[4]

Numerous studies have confirmed that extra-virgin olive oil is healthy to cook with. At worst, some of the oil's health benefits will become neutralized, but the fat is still very stable and damage-resistant. For higher temperatures, however, it's best to reach for fats with a higher level of saturation, meaning more likely to be solid at room temperature. These include butter, ghee, coconut oil, and beef tallow. As you begin to bring healthy fats back to the table, keep in mind that oils are calorie-rich, possessing more than double the calories per gram of protein or carbs, so don't go overboard! Use as needed to cook (1 tablespoon is usually adequate), and indulge in an optional 1 to 2 tablespoons of "raw" extra-virgin olive oil (120 to 240 calories), either as a dressing on your salad or added to eggs or veggies.

Always Foods

It's time to reload with the foods that are going to serve you. Many of these foods are Genius Foods, capable of arming your brain with the nutrients (like DHA fat) it needs not only to grow healthy new brain cells throughout your life, but to arm your brain with the defense artillery to protect against the numerous stresses it incurs along the way.

- **Oils and fats:** Extra-virgin olive oil, avocado oil, coconut oil, grass-fed tallow, organic or grass-fed butter and ghee.
- **Protein:** Grass-fed beef; free-range poultry; pasture-raised pork, lamb, bison, and elk; pastured or omega-3 eggs; wild salmon; sardines; anchovies; shellfish and mollusks (shrimp, crab, lobster, mussels, clams, oysters).
- **Nuts and seeds:** Almonds and almond butter, Brazil nuts, cashews, macadamias, pistachios, pecans, walnuts, flaxseeds, sunflower seeds, pumpkin seeds, sesame seeds, chia seeds.
- **Vegetables:** Mixed greens, kale, spinach, collard greens, mustard greens, broccoli, chard, cabbage, onions, mushrooms, cauliflower, Brussels sprouts, sauerkraut, kimchi, pickles,

artichokes, broccoli sprouts, green beans, celery, bok choy, watercress, asparagus, garlic, leeks, fennel, shallots, scallions, ginger, jicama, parsley, water chestnuts, nori, kelp, dulse seaweed.

▸ **Nonstarchy root vegetables:** Beets, carrots, radishes, turnips, parsnips.

▸ **Low-sugar fruits:** Avocados, coconut, olives, blueberries, blackberries, raspberries, grapefruit, kiwis, bell peppers, cucumbers, tomatoes, zucchini, squash, pumpkin, eggplant, lemons, limes, cacao nibs, okra.

▸ **Herbs, seasonings, and condiments:** Parsley, rosemary, thyme, cilantro, sage, turmeric, cinnamon, cumin, allspice, cardamom, ginger, cayenne, coriander, oregano, fenugreek, paprika, salt, black pepper, vinegar (apple cider, white, balsamic), mustard, horseradish, tapenade, salsa, nutritional yeast.

▸ **Fermented, organic soy:** Natto, miso, tempeh, organic gluten-free tamari sauce.

▸ **Dark chocolate:** At least 80% cocoa content (ideally 85% or higher).

▸ **Beverages:** Filtered water, coffee, tea, unsweetened almond/flax/coconut/cashew milk.

AVOID "SNACKCIDENTS" WITH PROTEIN

Remember from chapter 1 that protein can be a powerful tool in your fight for a better and more resilient body. It helps maintain and grow lean mass while also being highly satiating. As one prone to "snackcidents"—eating an excessive amount of low-quality calories from snacking—I understand the need for better snacks that are satiating while also feeling like a treat. Here are some good options:

▶ Low-sugar beef, turkey, or salmon jerky

▶ Biltong (an African style of jerky that is simply air-dried meat)

▶ Uncured sliced meats

▶ Full-fat or fat-free yogurt (see below)

▶ Pastured pork rinds with a sprinkle of nutritional yeast

▶ Hard-boiled eggs

Sometimes Foods

▶ **Dairy:** Grass-fed and antibiotic- and hormone-free yogurt and hard cheeses.

▶ **Legumes:** Beans, lentils, peas, chickpeas, hummus, peanuts.

▶ **Fiber extracts:** Products made with chicory root extract, tapioca fiber, soluble corn fiber, inulin. These ingredients are now used as sugar-free sweeteners and fiber sources. While they are likely fine in moderation, too much can cause gas and bloating, and the jury is still out on whether they all act as true, digestion-resistant fibers.

▶ **Sweeteners:** Stevia, non-GMO sugar alcohols (erythritol is best to use, followed by xylitol, which is naturally harvested from birch trees), monk fruit (luo han guo), allulose.

FAQ

Q: Should I go for full-fat or fat-free dairy?

A: Dairy is not essential for adults. But if you choose to enjoy it (assuming you're not one of the 75 percent of lactose-intolerant adults), here's the deal: people who consume full-fat dairy seem to be protected from the standpoint of heart and metabolic health, whereas no observable protection comes from low-fat or fat-free dairy.[5] Generally, if you're going to opt for dairy, choose full-fat, which comes with a number of beneficial compounds that are especially concentrated in dairy, like vitamin K_2 or the potential cancer fighter conjugated linoleic acid (CLA). That said, does low-fat or fat-free dairy have a place in the Genius Life? Yes! Fat-free Greek yogurt can be an excellent and highly satiating snack, with more protein and fewer calories than its full-fat counterpart. Just be sure to opt for plain varieties without added sugar, and add fruit or other toppings (berries, kiwi, and cacao nibs are personal favorites) yourself.

Prioritize Protein for Weight Loss and Better Health

This may surprise you: the order in which you consume foods can not only bolster your health but help you lose or maintain weight. In a given meal, try to prioritize protein, followed by fibrous vegetables, with your "carb" of choice last. Why does this trick work? Protein is highly satiating, and by eating your chicken (or steak, or fish) first, you ensure that your protein needs are met. Fibrous vegetables are next in line. Fiber absorbs water, expanding in your stomach. That mechanical stretching helps to turn off stomach-derived hormones that tell your brain "feed me." Later in the meal, starchier and less nutrient-dense foods like rice (or even the occasional dessert treat) can take what remains of your hunger. New research shows that by following this exact method (consuming carbs at the end of a meal), the resulting *post-meal* hunger is reduced.[6] In other words, you're going to be less likely to get hungry later. Skip the bread basket and have your carbs last!

Week 4: Determine Your Custom Carb Score

A two-week "induction period" of lower-carb, higher-protein eating serves two functions: it helps you to stock up on vital nutrition from nutrient-dense whole foods, and it allows you to regain metabolic flexibility, which is the ability to tap into your own body fat for fuel without experiencing excessive hunger. By dropping your insulin levels with a lower-carb diet and adding in some high-intensity exercise and resistance training, you become *fat-adapted*. The best part? After the second week, those concentrated sources of carbohydrates can be added back in to support your lifestyle.

How many servings of sweet potatoes and rice might you consume in a week? Unfortunately, there is no one-size-fits-all recommendation, but the following Carb Score continuum should help as a general guide to optimize your carbohydrate intake. Begin with zero and add the number of points for every box that you tick. The total number of points you end up with corresponds to how many servings from the following list of higher-carbohydrate foods you may consume per week.

Least tolerance (0–4 servings per week)		Highest tolerance (8–14 servings per week)	
Prediabetic or type 2 diabetic*	(0)	Neither prediabetic or type 2 diabetic	(+4)
Bulging waist	(0)	Slim waist	(+2)
Sedentary lifestyle	(0)	Active lifestyle	(+4)
Infrequent exercise	(0)	Regular (3–5x/week) resistance or high-intensity interval training	(+4)

---

* If you are a type 2 diabetic or prediabetic individual, you have developed carbohydrate intolerance, and should focus instead on the "Always Foods" listed earlier. If you take any type of medication for diabetes, consult with your doctor as sudden carb restriction could require you to adjust your medication, or risk a dangerous condition called diabetic ketoacidosis.

If your Carb Score was a 10, this means that you may have up to 10 servings from the list below every week. Ultimately, self-experimentation, sustainability, and how you feel will be the primary determinant. If you feel frequent hunger, you can try reducing your carbohydrate intake, and if your hunger is well controlled, you can try adding in more.

Now that you know your Carb Score, you may enjoy the following higher-carbohydrate foods. Keep in mind that a serving size of whole fruit is a 6-ounce portion, which equates roughly to a whole fruit in the case of an apple or orange. And the ideal time to consume them is during the day, or after a workout.

▸ **Starchy root vegetables:** White potatoes, sweet potatoes.
▸ **Non-gluten-containing unprocessed grains:** Buckwheat, rice (brown, white, wild), millet, quinoa, sorghum, teff, gluten-free oatmeal, non-GMO corn or popcorn. Oats do not naturally contain gluten but are frequently contaminated with gluten, as they are processed in facilities that also handle wheat. Therefore, look for oats that explicitly indicate on the package that they are gluten-free.
▸ **Whole, sweet fruit:** Apples, apricots, mangos, melons, pineapple, pomegranate, and bananas provide various nutrients and different types of fiber. Be cautious with dried fruit, which has the water removed and sugar concentrated, making it easy to overdo.

WHY I RECOMMEND LIMITED GRAINS

If you carry excessive weight, especially around your midsection (aka visceral fat), it's reasonable to assume that you have some degree of insulin resistance and thus poor control over your blood sugar. If so, you should probably go grain-free, sticking instead to fibrous veggies and protein. Keep in mind that there is no human requirement for grains, and grains do not provide any nutrient that cannot be obtained more easily (either in higher concentration or in a more bioavailable form) from other sources. If you don't have visceral fat or are exercising regularly, don't be afraid to consume that organic corn on the cob or small bowl of rice. Just be aware that protein and vegetables (particularly the nonstarchy variety) are *always* going to be the more nutrient-dense option.

24 Hours of Genius

Waking Up

A Genius morning involves waking up naturally, without an alarm clock. Yes, that is impractical for most people, but going to sleep earlier, if a possibility, may help. If you must wake up at a set time, try to use an alarm clock that eases the transition to wakefulness. I suggest the app Sleep Cycle for this. I have no affiliation with the company, but I do appreciate that it uses your smartphone's microphone to determine when you have entered a lighter phase of sleep prior to waking you up. The only downside is that using this app requires you to keep your phone by your bed, which increases the temptation that you're going to use it when you should be sleeping. For this reason, I recommend putting your phone in Airplane mode prior to going to sleep.

Morning

Now that you've woken up, your first priority—after peeing, perhaps—should be to have a glass of water. Eight ounces is fine. I will often put a dash of mineral salt in my water to replenish electrolytes. You become dehydrated during sleep; to what degree depends on various factors including room humidity, temperature, and whether or not you sweat at night. Also, lower-carbohydrate diets tend to lead to sodium depletion. Adding a pinch of a high-quality salt to your water can make you feel great, especially if you tend to feel light-headed upon waking.

FAQ

Q: How much water should I drink daily?

A: Even mild dehydration can lead to reduced cognitive function and mood, so it's important to stay hydrated.[7] There is no one-size-fits-all approach to how much water you need to drink to achieve that goal, though a general guideline is to have a cup upon waking and continue to hydrate throughout the day to ensure that your pee is clear or light yellow at the darkest. You can also reduce your need for drinking water by consuming water-based foods like soups, broths, and decaffeinated teas or by eating fruits and vegetables, which provide significant water. Activities that increase your need for fluids include caffeine consumption as well as any activity that makes you sweat (you can also lose water without sweating, which can occur in very dry climates).

After hydrating, the next step is circadian entrainment, which you learned the importance of in chapter 2. You can anchor your brain's twenty-four-hour clock by exposing your eyes to ambient sunlight. If you don't have a terrace or backyard, or if the weather doesn't permit, large open windows will do just fine. Try to spend at least a half hour of your morning in this naturally lit environment.

Instead of checking your phone during this period, take a moment to practice some stretching, deep breathing, or meditation. If you choose to meditate, do so *before* any coffee or other caffeinated beverage is consumed. A morning meditation is a great way to clear your head, draw your awareness into the present, and set your intention for the day to come. Revisit page 192 for a simple practice you can use.

Now that some time has passed since waking up, reach for that cup of coffee if you are so inclined! This delay is actually purposeful, since soon after rising comes your natural peak in cortisol, the hormone that provides morning wakefulness. As part of your natural circadian rhythm, it subsides a half hour to forty-five minutes after waking, at which point coffee becomes a delightful addition.

COFFEE: HEALTHY OR UNHEALTHY?

Coffee contains caffeine which, even at small doses, can give you a mental edge. It can also boost physical performance and strength. However, too much of a good thing can become a bad thing. Caffeine works by blocking chemicals in the brain that would otherwise make you feel fatigue. But coffee doesn't create energy from thin air; it borrows it from later. Over time, dependency can *lower* your performance, with every new cup merely treating the withdrawal we feel from the last.

To be clear, coffee can be good for you. The literature on coffee seems to be weighed in its favor, at least for the general population. In one large study, people who consumed one cup a day (even decaf) were 12 percent less likely to die from heart disease, cancer, stroke, diabetes, and respiratory and kidney disease over sixteen years compared to those who didn't drink any, and those who drank two to three cups per day were 18 percent less likely.[8] However, we all have different caffeine tolerances, dictated by stress and genes, among other things. Plus, our coffees are now getting larger and stronger, with a 16-ounce cold brew providing a whopping 200 milligrams of caffeine—double what we'd expect from a cup of home-brewed coffee. The take-home? Not everyone is going to feel great from coffee.

If you find yourself caught in a vicious loop of caffeine dependency and overuse (which usually comes with the telltale *wired-and-tired* feeling), try taking a week or two off every few months to resensitize yourself. (For the first few days you can use decaf as a weaning-off tool.) After an initial three-to-four-day hump, you likely won't even need coffee anymore. If and when you decide to go back, try to keep your consumption at the minimal effective dose—one small cup as needed—and consider a weekly "off" day or two to keep your coffee working for you, *not* the other way around.

Many people enjoy exercise in the morning; you either can work out or not—it's up to you! I enjoy morning workout sessions, but remember that time of day makes no significant difference on the benefits incurred from exercise. Regardless of where in your day you choose to fit your exercise, ensure that you achieve a blend of resistance training and high-intensity interval training. And if you've been (or plan to be) mostly sedentary all day, it doesn't hurt to throw in some cardio, as well. Revisit chapter 4 for a breakdown of each type of exercise with routines.

THE TERRIBLE CASE OF THE TOO-TALL TOILET

Make sure your daily toilet is low to the ground or you could be in for a digestive disaster. One day and without prior warning, the building manager of my Los Angeles rental installed a new Americans with Disabilities Act–compliant toilet in my home, which was much taller than what had been there previously (these toilets are perhaps easier to get on and off of, but are terrible for elimination). After the installation, my digestion was way off; I never felt like I was eliminating completely, and I was perpetually bloated. The worst part? I couldn't understand why; my diet hadn't changed, after all. Months later, it dawned on me: it was the toilet!

Humans have squatted to eliminate for the vast majority of history, and at least 1.2 billion people around the world still do. I had learned the hard way that sitting on a high toilet constrains the puborectalis muscle, which wraps around the rectum. This causes a shortening of the anorectal angle, which should be mostly straight during elimination. If your toilet is tall, do what I did; invest in a toilet stool to rest your feet on, thus allowing your knees to rise above your hips. You might be amazed at what a twenty-five-dollar investment can do for your digestion, health, and, yes, even your mood!

Late Morning

Late morning is a great time to have your first meal of the day, though you can enjoy it at any time that works best for you. As a general guideline, aim to have your first meal one to three hours after you wake up, and stop eating two to three hours before bed. Think of this first meal as setting your food intention for the day.

When you decide to eat, construct your plate with half your protein of choice and half vegetables. You may choose to have a large, colorful "fatty" salad—one of my go-to meals ("fatty" implies that it is accompanied by a healthy serving of extra-virgin olive oil, which aids nutrient absorption). Or have your protein with cooked veggies. Aim for variety and mix it up! You can't go wrong with this template, which is destined to satiate *and* provide ample nutrition.

EXAMPLE FIRST MEALS

Grass-fed beef (6 oz.) Dark leafy greens (10 oz.) Dressing: 1–2 tablespoons extra-virgin olive oil 1–2 tablespoons balsamic vinegar Salt Pepper Garlic	Chicken breast (6 oz.) Sautéed broccoli (10 oz.) Topping: Lemon juice Extra-virgin olive oil Chile flakes Mustard seed powder Salt	Eggs (3, whole, medium) Scrambled with: Spinach (2 oz.) Chopped bell peppers (2 oz.) ½ avocado Topped with: 1 tablespoon extra-virgin olive oil 2 tablespoons pico de gallo

Once you've found a routine that works for you, stick to it. Research suggests that skipping breakfast can lead to less efficient glucose handling, but *only* for those who regularly eat breakfast.[9] This implies that whether your first meal is at 9 a.m. or noon, aim to be more or less consistent.

Remember that the Genius Life incorporates the latest research on circadian biology, which stipulates that food should be consumed during a biologically appropriate time frame. In other words, eat during the day, and not too late at night. By constraining your feeding window to a set number of hours, a number of potential benefits may be imparted, including lower blood sugar, blood pressure, and oxidative stress, a driver of inflammation.[10] (For the snack-happy, this also has the added benefit of making it easier to consume fewer calories in general.)

Whenever you decide to eat, always be mindful and present with your food. Put away your smartphone, and don't eat with the television on. Studies show that when subjects are provided with a distraction—either printed material or a smartphone—they consume on average fifteen percent more calories.[11] Increasing your awareness of your food and eliminating distractions (including your smartphone, television, or even that juicy gossip magazine) could be a simple and deprivation-free means of cutting calories—and dropping weight.

Early Evening

Early evening is the ideal time to eat dinner, since it's likely when our ancestors would have gathered around the campfire to cook and tell stories (later in the night, as melatonin levels begin to rise, sleep becomes a greater priority). For dinner, arrive at the table and take a few slow, deep breaths to engage your parasympathetic nervous system. This is most commonly referred to as our "rest and digest" state, and it facilitates optimal nutrient absorption *and* minimizes the chances of digestive upset.

Dinner should be a hefty serving of protein and vegetables—feel free to experiment here. When eating satiating, nourishing foods, it is largely unnecessary to moderate portions or count calories, although doing so can be useful to push past a weight loss plateau. Focus on sautéed or roasted cruciferous vegetables like broccoli, cauliflower, and Brussels sprouts.

Many chefs like to focus on quality ingredients over quantity, and this is a philosophy I wholeheartedly endorse. It makes cooking delicious, healthy, Mediterranean-style meals easy and inexpensive! Here are some cooking ingredients you should always aim to have in your kitchen:

Mediterranean Ingredients for Easy Cooking

Eggs	Extra-virgin olive oil
Salt	Grass-fed butter
Pepper	Apple cider vinegar
Garlic powder	Lemon and lime
Mustard seed powder	Balsamic vinegar

With these simple ingredients, you can make any meat or vegetable savory and delicious. Keep in mind that extra-virgin olive oil is fine to use for low to medium temperature cooking, but for high heats you're better off using butter or ghee (the latter is clarified butter and is even more resistant to high-heat cooking). Feel free to keep a bottle of extra-virgin olive oil on the table to use as a sauce.

Eat slowly, chewing your food carefully; remember, digestion begins in the mouth. The mere act of chewing not only prepares stomach acid and enzymes to break down and absorb nutrients, but encourages the formation of unique and beneficial compounds in your food. One such compound, sulforaphane, is formed during the chewing of raw cruciferous vegetables (or with the addition of mustard seed powder to cooked veggies). Sulforaphane, mentioned on page 169, aids in the detoxification of various environmental pollutants.

Express gratitude for your food and, if consuming meat, the animal that sacrificed its life for your nourishment. And eat until fully satisfied, but no more. Be mindful that after your dinner is over, the "kitchen closes" and your feeding opportunity for the day is over as your body begins digestion and prepares for sleep. In other words, don't deprive yourself at dinner!

Going Out

Socializing is an important part of a Genius Life because social connection helps us to live longer, more fulfilling lives. (There's no point in living a Genius Life as a hermit!) Unfortunately, social obligations can often be at odds with health goals, usually involving peer pressure to drink alcohol or indulge in unhealthy foods. When planning social outings, try to pick health-forward restaurants or activities that don't require alcohol. Should you choose to imbibe, here are some simple steps to ensure minimal damage is done:

▶ **Always opt for spirits over beer.** Vodka or tequila are good options, and skip the sugary mixers (a dash of bitters or a lime are fine). Wine is also fine; just be sure to ask for a dry variety to cut down on sugar intake.

▶ **Drink a glass of water between every alcoholic drink.** If you start to notice that you are peeing a lot (alcohol can act like a diuretic), add some table salt and a squeeze of lime to your water to make an off-the-cuff electrolyte cocktail.

▶ **Drink on an empty stomach.** The conventional wisdom to not drink on an empty stomach is meant to blunt the impact of alcohol. But if you are chasing a buzz responsibly, your minimum effective dose will be smaller with less food in your stomach. This eases the processing burden on your liver, which will prioritize purging the toxin (alcohol) above other tasks— including properly digesting your food.

▶ **Sober up before bed.** Alcohol impairs sleep. Try to regain sobriety before hitting the hay to ensure you reap all of the reparative benefits of your slumber.

Above all, remember: moderation is key! This means one glass per day for women and one to two for men. This will allow you to reap the potential benefits of alcohol consumption (i.e., social lubrication, stress relief), while keeping the harm to an absolute minimum.

THE GENIUS MOCKTAIL

Who says abstinence from alcohol has to be boring? In 2019, I was in Bogotá, Colombia, when I discovered a phenomenal drink enjoyed by the locals. They called it a "sparkling water michelada." A michelada is typically a Mexican alcoholic drink comprising beer and tomato juice, served in a salted rim glass. The Colombian take on it, however, is alcohol-, tomato-, and calorie-free. Refreshing and healthy, making one is easy and inexpensive. And the best part is that it's just as fun as a cocktail to enjoy in a bar or restaurant setting!

WHAT YOU'LL NEED:

½-1 lime

12 ounces sparkling or seltzer water

6 ice cubes

2 tablespoons coarse salt

1 tall glass

1 plate

WHAT TO DO:

Place salt on the plate and distribute evenly.

Slice the lime and run it around the rim of the glass.

As if you were making a margarita, dip the glass, rim side down, into the salt. Coat evenly.

Fill the glass with ice.

Squeeze half a lime into the glass (more depending on the size of the glass).

Pour in sparkling water.

Feel free to drink this any time! It contains virtually zero calories but provides vitamin C from the lime and electrolytes from the salt. You'll look just as fancy as with a cocktail, but without the energy-zapping alcohol!

Before Bed

After dinner, begin priming your body for sleep. You may choose to watch a movie or TV show, or read a book. If your choice is the former, be sure to time the movie (or show) so that you can still unwind afterward. If you tend to watch up until bedtime, it can be helpful to wear your blue-blocking glasses to mitigate the light-induced melatonin suppression. These should be worn for two to three hours before the time you intend to go to sleep.

Final Notes

I hope the guidelines given throughout this book impart good health in both body and mind. Feel free to take the ideas and let them be jumping off points for your own research and experimentation. In health, there are few concrete answers because we all come with a different set of genes, habits, habitats, and predilections. In fact, the answer to most questions is "It depends!" Everyone's needs are different, and they change over time—your needs now are different from what they will be in ten years.

There's no denying the brokenness of our modern environment, and the specifics laid out in this book have the potential to radically improve your health and well-being. Science is continually evolving, and what we assume to be true today might be disproven in the future. Nonetheless, we must be proactive about our health. I was shocked to learn how limited the options were for each of my mother's freak medical conditions. It made me see the importance of doing what we can to remain healthy *before* such problems occur.

Our world has changed; our devices are here to stay. Artificial lighting and climate control are not going anywhere. Neither is plastic or processed foods. But by arming ourselves and our loved ones with knowledge, we can make decisions that may buy us additional days, months, or years of healthy time. You deserve better than to live sick and die young. You deserve a long and healthy life—a Genius Life. Make the effort to stay strong in your body and mind, eat the right foods, and be mindful of your surroundings; perhaps you'll enjoy a different fate than my mom and so many like her.

ACKNOWLEDGMENTS

Writing a book isn't easy, especially doing so in the midst of family tragedy. I'm grateful to have been surrounded by people who have provided the encouragement to keep my head up and press on. I'd like to thank first my agent, Giles Anderson, for being indispensable to the process, as well as my publisher, Karen Rinaldi, and everyone at Harper Wave for continually believing in my mission.

I'd like to thank my brothers, Ben and Andrew, and my dad, Bruce. Also my cat, Delilah. Blame her for any typos you find.

My brilliant and caring friends Sarah Anne Stewart and Craig Clemens. Love you guys both so much.

My good friend and Alzheimer's prevention pioneer Richard Isaacson. You are a legend. Thank you for all the inspiration and always being there for me to help hone my message.

My friends Mark Hyman, David Perlmutter, Dhru Purohit, and Andrew Luer.

Other Geniuses who have helped immensely: Kristin Loberg, Sal Di Stefano, Chris Masterjohn, Kate Adams, and Carol Kwiatkowski.

To every one of my followers on Instagram, Facebook, YouTube, and Twitter, and to listeners of *The Genius Life* podcast, thank you for the incredible support every step of the way.

To each backer of my documentary, *Bread Head*, thank you for your patience and believing in my ability to tell this story through film.

To all of the researchers whose work I cite, thank you for helping to shine a light on the mysteries of our incredibly elegant but vulnerable bodies and brains. I owe you a debt of gratitude for your work in the trenches.

And to you for buying my book. I couldn't have done this without your support, so thank you.

NOTES

PREFACE

1. Max Lugavere, Alon Seifan, and Richard S. Isaacson, "Prevention of Cognitive Decline," *Handbook on the Neuropsychology of Aging and Dementia* ed. Lisa Ravdin and Heather Katzen, (Springer, Cham, 2019), 205–29.

2. Hugh C. Hendrie et al., "APOE ε4 and the Risk for Alzheimer's Disease and Cognitive Decline in African Americans and Yoruba," *International Psychogeriatrics* 26.6 (2014): 977–85.

3. A. M. Noone et al., SEER Cancer Statistics Review, 1975–2015, National Cancer Institute, Bethesda, MD, https://seer.cancer.gov/csr/1975_2015/, based on November 2017 SEER data submission, posted to the SEER website, April 2018.

4. CDC Newsroom, "Cancers Associated with Overweight and Obesity Make Up 40 Percent of Cancers Diagnosed in the United States," Centers for Disease Control and Prevention, October 3, 2017, www.cdc.gov/media/releases/2017/p1003-vs-cancer-obesity.html.

5. Ashkan Afshin et al., "Health Effects of Dietary Risks in 195 Countries, 1990–2017: A Systematic Analysis for the Global Burden of Disease Study 2017," *Lancet* 393, no. 10184 (2019): 1958–72.

6. George DeMaagd and Ashok Philip, "Parkinson's Disease and Its Management: Part 1: Disease Entity, Risk Factors, Pathophysiology, Clinical Presentation, and Diagnosis," *P & T: A Peer-Reviewed Journal for Managed Care and Hospital Formulary Management* 40.8 (2015): 504–32.

INTRODUCTION

1. Joana Araújo, Jianwen Cai, and June Stevens, "Prevalence of Optimal Metabolic Health in American Adults: National Health and Nutrition Examination Survey 2009–2016," Metabolic Syndrome and Related Disorders 17, no. 1 (2019): 46–52.

2. Jeffrey Gassen et al., "Inflammation Predicts Decision-Making Characterized by Impulsivity, Present Focus, and an Inability to Delay Gratification," *Scientific Reports* 9 (2019); Leonie J.T. Balter et al., "Selective Effects of Acute Low-Grade Inflammation on Human Visual Attention." *NeuroImage* 202 (2019): 116098; Felger, Jennifer C. "Imaging the Role of Inflammation in Mood and Anxiety-Related Disorders." *Current Neuropharmacology* 16, no. 5 (2018): 533–558.

3. Ole Köhler-Forsberg et al., "Efficacy of Anti-Inflammatory Treatment on Major Depressive Disorder or Depressive Symptoms: Meta-Analysis of Clinical Trials," *Acta Psychiatrica Scandinavica* 139.5 (2019): 404–19.

1: DON'T FORK AROUND

1. Christopher D. Gardner et al., "Effect of Low-Fat vs. Low-Carbohydrate Diet on 12-Month Weight Loss in Overweight Adults and the Association with Genotype

Pattern or Insulin Secretion: The DIETFITS Randomized Clinical Trial," *JAMA* 319.7 (2018): 667–79.

2. Isaac Abel, "Was I Actually 'Addicted' to Internet Pornography?" *Atlantic*, June 7, 2013, www.theatlantic.com/health/archive/2013/06/was-i-actually-addicted-to-internet -pornography/276619/.

3. Kevin D. Hall et al., "Ultra-Processed Diets Cause Excess Calorie Intake and Weight Gain: An Inpatient Randomized Controlled Trial of Ad Libitum Food Intake," *Cell Metabolism* 30 (2019): 67–77.

4. Sadie B. Barr and Jonathan C. Wright, "Postprandial Energy Expenditure in Whole-Food and Processed-Food Meals: Implications for Daily Energy Expenditure," *Food & Nutrition Research* 54 (2010), doi:10.3402/fnr.v54i0.5144.

5. Gloria González-Saldivar et al., "Skin Manifestations of Insulin Resistance: From a Biochemical Stance to a Clinical Diagnosis and Management," *Dermatology and Therapy* 7.1 (2016): 37–51, doi:10.1007/s13555-016-0160-3.

6. W. J. Lossow and I. L. Chaikoff, "Carbohydrate Sparing of Fatty Acid Oxidation. I. The Relation of Fatty Acid Chain Length to the Degree of Sparing. II. The Mechanism by Which Carbohydrate Spares the Oxidation of Palmitic Acid," *Archives of Biochemistry and Biophysics* 57.1 (1955): 23–40.

7. Andrew A. Gibb and Bradford G. Hill, "Metabolic Coordination of Physiological and Pathological Cardiac Remodeling," *Circulation Research* 123.1 (2018): 107–28.

8. Deniz Senyilmaz-Tiebe et al., "Dietary Stearic Acid Regulates Mitochondria in Vivo in Humans," *Nature Communications* 9, no. 1 (2018): 3129.

9. P. W. Siri-Tarino et al., "Saturated Fat, Carbohydrate, and Cardiovascular Disease," *Americal Journal of Clinical Nutrition* 91, no. 3 (2010): 502–9, doi:10.3945/ajcn.2008 .26285.

10. Christopher E. Ramsden et al., "Re-evaluation of the Traditional Diet-Heart Hypothesis: Analysis of Recovered Data from Minnesota Coronary Experiment (1968–73)," *BMJ* 353 (2016): i1246.

11. Stephan J. Guyenet and Susan E. Carlson, "Increase in Adipose Tissue Linoleic Acid of US Adults in the Last Half Century," *Advances in Nutrition* 6, no. 6 (2015): 660–64.

12. Manish Mittal et al., "Reactive Oxygen Species in Inflammation and Tissue Injury," *Antioxidants & Redox Signaling* 20.7 (2014): 1126–67.

13. Karen S. Bishop et al., "An Investigation into the Association Between DNA Damage and Dietary Fatty Acid in Men with Prostate Cancer," *Nutrients* 7, no. 1 (2015): 405–22, doi:10.3390/nu7010405.

14. H. Lodish et al., *Molecular Cell Biology*, 4th edition (New York: W. H. Freeman, 2000), section 12.4, "DNA Damage and Repair and Their Role in Carcinogenesis," available from https://www.ncbi.nlm.nih.gov/books/NBK21554/.

15. Shosuke Kawanishi et al., "Crosstalk Between DNA Damage and Inflammation in the Multiple Steps of Carcinogenesis," *International Journal of Molecular Sciences* 18, no. 8 (2017): 1808, doi:10.3390/ijms18081808.

16. Bruce N. Ames, "Prolonging Healthy Aging: Longevity Vitamins and Proteins," *Proceedings of the National Academy of Sciences* 115, no. 43 (2018): 10836–44.

17. Somdat Mahabir et al., "Dietary Magnesium and DNA Repair Capacity as Risk Factors for Lung Cancer," *Carcinogenesis* 29, no. 5 (2008): 949–56.

18. Takanori Honda et al., "Serum Elaidic Acid Concentration and Risk of Dementia: The Hisayama Study," *Neurology* (2019).

19. Jessica E. Saraceni, "8,000-Year-Old Olive Oil Found in Israel," *Archaeology*, www.archaeology.org/news/2833–141217-israel-galilee-olive-oil.

20. Felice N. Jacka et al., "A Randomised Controlled Trial of Dietary Improvement for Adults with Major Depression (the 'SMILES' trial)," *BMC Medicine* 15, no. 1 (2017): 23.

21. Marta Czarnowska and Elzbieta Gujska, "Effect of Freezing Technology and Storage Conditions on Folate Content in Selected Vegetables," *Plant Foods for Human Nutrition* 67, no. 4 (2012): 401–6.

22. Kristen L. Nowak et al., "Serum Sodium and Cognition in Older Community-Dwelling Men," *Clinical Journal of the American Society of Nephrology* 1, no. 3 (2018): 366–74.

23. Andrew Mente et al., "Urinary Sodium Excretion, Blood Pressure, Cardiovascular Disease, and Mortality: A Community-Level Prospective Epidemiological Cohort Study," *Lancet* 392, no. 10146 (2018): 496–506.

24. Loren Cordain et al., "Origins and Evolution of the Western Diet: Health Implications for the 21st Century," *American Journal of Clinical Nutrition* 81, no. 2 (2005): 341–54.

25. Robert R. Wolfe et al., "Optimizing Protein Intake in Adults: Interpretation and Application of the Recommended Dietary Allowance Compared with the Acceptable Macronutrient Distribution Range," *Advances in Nutrition* 8, no. 2 (2017): 266–75, doi: 10.3945/an.116.013821.

26. Robert W. Morton et al., "A Systematic Review, Meta-Analysis and Meta-Regression of the Effect of Protein Supplementation on Resistance Training-Induced Gains in Muscle Mass and Strength in Healthy Adults," *British Journal of Sports Medicine* 52, no. 6 (2017): 376–84, doi:10.1136/bjsports-2017-097608.

27. Michaela C. Devries et al., "Changes in Kidney Function Do Not Differ Between Healthy Adults Consuming Higher- Compared with Lower- or Normal-Protein Diets: A Systematic Review and Meta-Analysis," *Journal of Nutrition* 148, no. 11 (2018): 1760–75, doi:10.1093/jn/nxy197.

28. Stuart M. Phillips, Stéphanie Chevalier, and Heather J. Leidy, "Protein 'Requirements' Beyond the RDA: Implications for Optimizing Health," *Applied Physiology, Nutrition, and Metabolism* 41, no. 5 (2016): 565–72.

29. Claudia Martinez-Cordero et al., "Testing the Protein Leverage Hypothesis in a Free-Living Human Population," *Appetite* 59, no. 2 (2012): 312–15.

30. David S. Weigle et al., "A High-Protein Diet Induces Sustained Reductions in Appetite, Ad Libitum Caloric Intake, and Body Weight Despite Compensatory Changes in Diurnal Plasma Leptin and Ghrelin Concentrations," *American Journal of Clinical Nutrition* 82.1 (2005): 41–48; S. J. Long, A. R. Jeffcoat, and D. J. Millward, "Effect of Habitual Dietary-Protein Intake on Appetite and Satiety," *Appetite* 35, no. 1 (2000): 79–88.

31. Klaas R. Westerterp, "Diet Induced Thermogenesis," *Nutrition & Metabolism* 1, no. 1 (2004): 5, doi:10.1186/1743-7075-1-5.

32. Claire Fromentin et al., "Dietary Proteins Contribute Little to Glucose Production, Even Under Optimal Gluconeogenic Conditions in Healthy Humans," *Diabetes* 62, no. 5 (2013): 1435–42, doi:10.2337/db12-1208.

33. W. M. A. D. Fernando et al., "Associations of Dietary Protein and Fiber Intake with Brain and Blood Amyloid-β," *Journal of Alzheimer's Disease* 61, no. 4 (2018): 1589–98.

34. Joel Brind et al., "Dietary Glycine Supplementation Mimics Lifespan Extension by Dietary Methionine Restriction in Fisher 344 Rats," *FASEB Journal* 25, no. 1 (2011).

35. Richard A. Miller, et al. "Glycine Supplementation Extends Lifespan of Male and Female Mice," *Aging Cell* 18.3 (2019): e12953.

36. Enrique Meléndez-Hevia et al., "A Weak Link in Metabolism: The Metabolic Capacity for Glycine Biosynthesis Does Not Satisfy the Need for Collagen Synthesis," *Journal of Biosciences* 34, no. 6 (2009): 853–72.

37. Joseph Firth et al., "The Effects of Dietary Improvement on Symptoms of Depression and Anxiety: A Meta-Analysis of Randomized Controlled Trials," *Psychosomatic Medicine* 81, no. 3 (2019): 265–80, doi:10.1097/PSY.0000000000000673.

38. Donald R. Davis, Melvin D. Epp, and Hugh D. Riordan, "Changes in USDA Food Composition Data for 43 Garden Crops, 1950 to 1999," *Journal of the American College of Nutrition* 23, no. 6 (2004): 669–82.

39. Irakli Loladze, "Hidden Shift of the Ionome of Plants Exposed to Elevated CO_2 Depletes Minerals at the Base of Human Nutrition," *eLife* 3 (2014): e02245, doi:10.7554/eLife.02245.

40. Donald R. Davis, "Trade-Offs in Agriculture and Nutrition," *Food Technology* 59, no. 3 (2005): 120.

41. Marcin Baranski et al., "Higher Antioxidant Concentrations, and Less Cadmium and Pesticide Residues in Organically Grown Crops: A Systematic Literature Review and Meta-Analyses," *British Journal of Nutrition* 5, no. 112 (2014): 794–811.

42. Zhi-Yong Zhang, Xian-Jin Liu, and Xiao-Yue Hong, "Effects of Home Preparation on Pesticide Residues in Cabbage," *Food Control* 18, no. 12 (2007): 1484–87; Tianxi Yang et al., "Effectiveness of Commercial and Homemade Washing Agents in Removing Pesticide Residues on and in Apples," *Journal of Agricultural and Food Chemistry* 65, no. 44 (2017): 9744–52.

43. Martha Clare Morris et al., "Nutrients and Bioactives in Green Leafy Vegetables and Cognitive Decline: Prospective Study," *Neurology* 90, no. 3 (2018): e214–22.

44. Emily R. Bovier and Billy R. Hammond, "A Randomized Placebo-Controlled Study on the Effects of Lutein and Zeaxanthin on Visual Processing Speed in Young Healthy Subjects," *Archives of Biochemistry and Biophysics* 572 (2015): 54-57; Lisa M. Renzi-Hammond et al., "Effects of a Lutein and Zeaxanthin Intervention on Cognitive Function: A Randomized, Double-Masked, Placebo-Controlled Trial of Younger Healthy Adults." *Nutrients* 9.11 (2017): 1246, doi:10.3390/nu9111246.

45. Marcia C. de Oliveira Otto et al., "Everything in Moderation—Dietary Diversity and Quality, Central Obesity and Risk of Diabetes," *PLOS ONE* 10, no. 10 (2015): e0141341.

46. Bernard P. Kok et al., "Intestinal Bitter Taste Receptor Activation Alters Hormone Secretion and Imparts Metabolic Benefits," *Molecular Metabolism* 16 (2018): 76–87, doi:10.1016/j.molmet.2018.07.013.

2: TIMING IS EVERYTHING

1. Valter D. Longo and Satchidananda Panda, "Fasting, Circadian Rhythms, and Time-Restricted Feeding in Healthy Lifespan," *Cell Metabolism* 23, no. 6 (2016): 1048–59, doi:10.1016/j.cmet.2016.06.001.

2. Patricia L. Turner and Martin A. Mainster, "Circadian Photoreception: Ageing and the Eye's Important Role in Systemic Health," *British Journal of Ophthalmology* 92, no. 11 (2008): 1439–44.

3. Neil E. Klepeis et al., "The National Human Activity Pattern Survey (NHAPS): A Resource for Assessing Exposure to Environmental Pollutants," *Journal of Exposure Science and Environmental Epidemiology* 11, no. 3 (2001): 231.

4. David Montaigne et al., "Daytime Variation of Perioperative Myocardial Injury in

Cardiac Surgery and Its Prevention by Rev-Erb Antagonism: A Single-Centre Propensity-Matched Cohort Study and a Randomised Study," *Lancet* 391, no. 10115 (2018): 59–69.

5. Fariba Raygan et al., "Melatonin Administration Lowers Biomarkers of Oxidative Stress and Cardio-Metabolic Risk in Type 2 Diabetic Patients with Coronary Heart Disease: A Randomized, Double-Blind, Placebo-Controlled Trial," *Clinical Nutrition* 38, no. 1 (2017): 191–96.

6. D. X. Tan et al., "Significance and Application of Melatonin in the Regulation of Brown Adipose Tissue Metabolism: Relation to Human Obesity," *Obesity Reviews* 12, no. 3 (2011): 167–88.

7. Ran Liu et al., "Melatonin Enhances DNA Repair Capacity Possibly by Affecting Genes Involved in DNA Damage Responsive Pathways," *BMC Cell Biology* 14, no. 1 (2013): 1.

8. Leonard A. Sauer, Robert T. Dauchy, and David E. Blask, "Polyunsaturated Fatty Acids, Melatonin, and Cancer Prevention," *Biochemical Pharmacology* 61, no. 12 (2001): 1455–62.

9. M. Nathaniel Mead, "Benefits of Sunlight: A Bright Spot for Human Health," *Environmental Health Perspectives* 116, no. 4 (2008): A160–67, doi:10.1289/ehp.116-a160.

10. Tina M. Burke et al., "Effects of Caffeine on the Human Circadian Clock in Vivo and in Vitro," *Science Translational Medicine* 7, no. 305 (2015): 305ra146–305ra146.

11. Lisa A. Ostrin, Kaleb S. Abbott, and Hope M. Queener, "Attenuation of Short Wavelengths Alters Sleep and the ipRGC Pupil Response," *Ophthalmic and Physiological Optics* 37, no. 4 (2017): 440–50.

12. James Stringham, Nicole Stringham, and Kevin O'Brien, "Macular Carotenoid Supplementation Improves Visual Performance, Sleep Quality, and Adverse Physical Symptoms in Those with High Screen Time Exposure," *Foods* 6, no. 7 (2017): 47.

13. Shawn D. Youngstedt, Jeffrey A. Elliott, and Daniel F. Kripke, "Human Circadian Phase-Response Curves for Exercise," *Journal of Physiology* 597, no. 8 (2019): 2253–68.

14. Katri Peuhkuri, Nora Sihvola, and Riitta Korpela, "Dietary Factors and Fluctuating Levels of Melatonin," *Food & Nutrition Research* 56, no. 1 (2012): 17252.

15. Kazunori Ohkawara et al., "Effects of Increased Meal Frequency on Fat Oxidation and Perceived Hunger," *Obesity* 21.2 (2013): 336–43; Hana Kahleova et al., "Meal Frequency and Timing Are Associated with Changes in Body Mass Index in Adventist Health Study 2," *Journal of Nutrition* 147, no. 9 (2017): 1722–28.

16. Eve Van Cauter, Kenneth S. Polonsky, and André J. Scheen, "Roles of Circadian Rhythmicity and Sleep in Human Glucose Regulation," *Endocrine Reviews* 18, no. 5 (1997): 716–38.

17. Frank A. J. L. Scheer et al., "Adverse Metabolic and Cardiovascular Consequences of Circadian Misalignment," *Proceedings of the National Academy of Sciences* 106, no. 11 (2009): 4453–58; Yukie Tsuchida, Sawa Hata, and Yoshiaki Sone, "Effects of a Late Supper on Digestion and the Absorption of Dietary Carbohydrates in the Following Morning," *Journal of Physiological Anthropology* 32, no. 1 (2013): 9.

18. Megumi Hatori et al., "Time-Restricted Feeding Without Reducing Caloric Intake Prevents Metabolic Diseases in Mice Fed a High-Fat Diet," *Cell Metabolism* 15, no. 6 (2012): 848–60.

19. Kelsey Gabel et al., "Effects of 8-Hour Time-Restricted Feeding on Body Weight and Metabolic Disease Risk Factors in Obese Adults: A Pilot Study," *Nutrition and Healthy Aging* preprint (2018): 1–9; Elizabeth F. Sutton et al., "Early Time-Restricted

Feeding Improves Insulin Sensitivity, Blood Pressure, and Oxidative Stress Even Without Weight Loss in Men with Prediabetes," *Cell Metabolism* 27, no. 6 (2018): 1212–21.

20. Manolis Kogevinas et al., "Effect of Mistimed Eating Patterns on Breast and Prostate Cancer Risk (MCC-Spain Study)," *International Journal of Cancer* 143, no. 10 (2018): 2380–89.

21. Catherine R. Marinac et al., "Prolonged Nightly Fasting and Breast Cancer Prognosis," *JAMA Oncology* 2, no. 8 (2016): 1049–55.

22. Patricia Rubio-Sastre et al., "Acute Melatonin Administration in Humans Impairs Glucose Tolerance in Both the Morning and Evening," *Sleep* 37, no. 10 (2014): 1715–19.

23. David Lehigh Allen et al., "Acute Daily Psychological Stress Causes Increased Atrophic Gene Expression and Myostatin-Dependent Muscle Atrophy," *American Journal of Physiology–Heart and Circulatory Physiology* 299, no. 3 (2010): R889–98.

24. Javier T. Gonzalez et al., "Breakfast and Exercise Contingently Affect Postprandial Metabolism and Energy Balance in Physically Active Males," *British Journal of Nutrition* 110, no. 4 (2013): 721–32.

25. Elizabeth A. Thomas et al., "Usual Breakfast Eating Habits Affect Response to Breakfast Skipping in Overweight Women," *Obesity* 23, no. 4 (2015): 750–59, doi:10.1002/oby.21049.

26. Ricki J. Colman et al., "Caloric Restriction Reduces Age-Related and All-Cause Mortality in Rhesus Monkeys," *Nature Communications* 5 (2014): 3557.

27. Rai Ajit K. Srivastava et al., "AMP-Activated Protein Kinase: An Emerging Drug Target to Regulate Imbalances in Lipid and Carbohydrate Metabolism to Treat Cardio-Metabolic Diseases," Thematic Review Series: New Lipid and Lipoprotein Targets for the Treatment of Cardiometabolic Diseases, *Journal of Lipid Research* 53 no. 12 (2012): 2490–514.

28. Belinda Seto, "Rapamycin and mTOR: A Serendipitous Discovery and Implications for Breast Cancer," *Clinical and Translational Medicine* 1, no. 1 (2012): 29.

29. Francesca LiCausi and Nathaniel W. Hartman, "Role of mTOR Complexes in Neurogenesis," *International Journal of Molecular Sciences* 19, no. 5 (2018): 1544, doi:10.3390/ijms19051544.

30. Alessandro Bitto et al., "Transient Rapamycin Treatment Can Increase Lifespan and Healthspan in Middle-Aged Mice," *eLife* 5 (2016): e16351.

31. Sebastian Brandhorst et al., "A Periodic Diet That Mimics Fasting Promotes Multi-System Regeneration, Enhanced Cognitive Performance, and Healthspan," *Cell Metabolism* 22, no. 1 (2015): 86–99, doi:10.1016/j.cmet.2015.05.012.

32. Ibid.

33. Sushil Kumar and Gurcharan Kaur, "Intermittent Fasting Dietary Restriction Regimen Negatively Influences Reproduction in Young Rats: A Study of Hypothalamo-Hypophysial-Gonadal Axis," *PLOS ONE* 8, no. 1 (2013): e52416.

3: THE VIGOR TRIGGER

1. Thomas J. Littlejohns et al., "Vitamin D and the Risk of Dementia and Alzheimer Disease," *Neurology* 83, no. 10 (2014): 920–28.

2. Lewis O. J. Killin et al., "Environmental Risk Factors for Dementia: A Systematic Review," *BMC Geriatrics* 16, no. 1 (2016): 175.

3. Joshua W. Miller et al., "Vitamin D Status and Rates of Cognitive Decline in a Multi-ethnic Cohort of Older Adults," *JAMA Neurology* 72, no. 11 (2015): 1295–303.

4. Jingya Jia et al. "Effects of Vitamin D Supplementation on Cognitive Function and Blood A -Related Biomarkers in Older Adults with Alzheimer's Disease: A Randomised, Double-Blind, Placebo-Controlled Trial." *Journal of Neurology, Neurosurgery & Psychiatry* (2019): jnnp-2018.

5. Robert Briggs et al., "Vitamin D Deficiency Is Associated with an Increased Likelihood of Incident Depression in Community-Dwelling Older Adults," *Journal of the American Medical Directors Association* 20, no. 5 (2019): 517–23.

6. Daniel A. Nation et al., "Blood–Brain Barrier Breakdown Is an Early Biomarker of Human Cognitive Dysfunction," *Nature Medicine* 25, no. 2 (2019): 270–76.

7. Peter Brøndum-Jacobsen et al., "25-hydroxyvitamin D and Symptomatic Ischemic Stroke: An Original Study and Meta-Analysis," *Annals of Neurology* 73, no. 1 (2013): 38–47.

8. Pauline Maillard et al., "Effects of Arterial Stiffness on Brain Integrity in Young Adults from the Framingham Heart Study," *Stroke* 47, no. 4 (2016): 1030–36; Joel Singer et al., "Arterial Stiffness, the Brain and Cognition: A Systematic Review," *Ageing Research Reviews* 15 (2014): 16–27.

9. Angela L. Jefferson et al., "Higher Aortic Stiffness Is Related to Lower Cerebral Blood Flow and Preserved Cerebrovascular Reactivity in Older Adults," *Circulation* 138, no. 18 (2018): 1951–62.

10. Noel T. Mueller et al., "Association of Age with Blood Pressure Across the Lifespan in Isolated Yanomami and Yekwana Villages," *JAMA Cardiology* 3, no. 12 (2018): 1247–49.

11. Daniel Lemogoum et al., "Effects of Hunter-Gatherer Subsistence Mode on Arterial Distensibility in Cameroonian Pygmies," *Hypertension* 60, no. 1 (2012): 123–28.

12. Ibhar Al Mheid et al., "Vitamin D Status Is Associated with Arterial Stiffness and Vascular Dysfunction in Healthy Humans," *Journal of the American College of Cardiology* 58, no. 2 (2011): 186–92.

13. Cedric F. Garland et al., "Meta-Analysis of All-Cause Mortality According to Serum 25-hydroxyvitamin D," *American Journal of Public Health* 104, no. 8 (2014): e43–50; Jacqueline A. Pettersen, "Vitamin D and Executive Functioning: Are Higher Levels Better?" *Journal of Clinical and Experimental Neuropsychology* 38, no. 4 (2016): 467–77.

14. Heike A. Bischoff-Ferrari et al., "Estimation of Optimal Serum Concentrations of 25-hydroxyvitamin D for Multiple Health Outcomes," *American Journal of Clinical Nutrition* 84, no. 1 (2006): 18–28.

15. John Paul Ekwaru et al., "The Importance of Body Weight for the Dose Response Relationship of Oral Vitamin D Supplementation and Serum 25-hydroxyvitamin D in Healthy Volunteers," *PLOS ONE* 9, no. 11 (2014): e111265.

16. Anas Raed et al., "Dose Responses of Vitamin D_3 Supplementation on Arterial Stiffness in Overweight African Americans with Vitamin D Deficiency: A Placebo Controlled Randomized Trial," *PLOS ONE* 12, no. 12 (2017): e0188424.

17. Donald Liu et al., "UVA Irradiation of Human Skin Vasodilates Arterial Vasculature and Lowers Blood Pressure Independently of Nitric Oxide Synthase," *Journal of Investigative Dermatology* 134, no. 7 (2014): 1839–46.

18. Yong Zhang et al., "Vitamin D Inhibits Monocyte/Macrophage Proinflammatory Cytokine Production by Targeting MAPK Phosphatase-1," *Journal of Immunology* 188, no. 5 (2012): 2127–35.

19. Kai Yin and Devendra K. Agrawal, "Vitamin D and Inflammatory Diseases," *Journal of Inflammation Research* 7 (2014): 69.

20. JoAnn E. Manson et al., "Vitamin D Supplements and Prevention of Cancer and Cardiovascular Disease," *New England Journal of Medicine* 380, no. 1 (2019): 33–44.

21. Aaron Lerner, Patricia Jeremias, and Torsten Matthias, "The World Incidence and Prevalence of Autoimmune Diseases Is Increasing," *International Journal of Celiac Disease* 3, no. 4 (2015): 151–55.

22. Wendy Dankers et al., "Vitamin D in Autoimmunity: Molecular Mechanisms and Therapeutic Potential," *Frontiers in Immunology* 697, no. 7 (2017), doi:10.3389/fimmu.2016.00697.

23. Ruth Dobson, Gavin Giovannoni, and Sreeram Ramagopalan, "The Month of Birth Effect in Multiple Sclerosis: Systematic Review, Meta-Analysis and Effect of Latitude," *Journal of Neurology, Neurosurgery, and Psychiatry* 84, no. 4 (2013): 427–32.

24. Emily Evans, Laura Piccio, and Anne H. Cross, "Use of Vitamins and Dietary Supplements by Patients with Multiple Sclerosis: A Review," *JAMA Neurology* 75, no. 8 (2018): 1013–21.

25. Barbara Prietl et al., "Vitamin D Supplementation and Regulatory T Cells in Apparently Healthy Subjects: Vitamin D Treatment for Autoimmune Diseases?" *Israel Medical Association Journal: IMAJ* 12, no. 3 (2010): 136–39.

26. Tara Raftery et al., "Effects of Vitamin D Supplementation on Intestinal Permeability, Cathelicidin and Disease Markers in Crohn's Disease: Results from a Randomised Double-Blind Placebo-Controlled Study," *United European Gastroenterology Journal* 3, no. 3 (2015): 294–302.

27. Danilo C. Finamor et al., "A Pilot Study Assessing the Effect of Prolonged Administration of High Daily Doses of Vitamin D on the Clinical Course of Vitiligo and Psoriasis," *Dermato-Endocrinology* 5, no. 1 (2013): 222–34, doi:10.4161/derm.24808.

28. Yasumichi Arai et al., "Inflammation, but Not Telomere Length, Predicts Successful Ageing at Extreme Old Age: A Longitudinal Study of Semi-Supercentenarians," *EBioMedicine* 2, no. 10 (2015): 1549–58.

29. Adam Kaplin and Laura Anzaldi, "New Movement in Neuroscience: A Purpose-Driven Life," *Cerebrum: The Dana Forum on Brain Science*, Dana Foundation, vol. 2015.

30. J. Brent Richards et al., "Higher Serum Vitamin D Concentrations Are Associated with Longer Leukocyte Telomere Length in Women," *American Journal of Clinical Nutrition* 86, no. 5 (2007): 1420–25, doi:10.1093/ajcn/86.5.1420.

31. Karla A. Mark et al., "Vitamin D Promotes Protein Homeostasis and Longevity via the Stress Response Pathway Genes skn-1, ire-1, and xbp-1," *Cell Reports* 17, no. 5 (2016): 1227–37.

32. Angela Carrelli et al., "Vitamin D Storage in Adipose Tissue of Obese and Normal Weight Women," *Journal of Bone and Mineral Research* 32, no. 2 (2016): 237–42, doi: 10.1002/jbmr.2979.

33. John Paul Ekwaru et al., "The Importance of Body Weight for the Dose Response Relationship of Oral Vitamin D Supplementation and Serum 25-hydroxyvitamin D in Healthy Volunteers," *PLOS ONE* 9, no. 11 (2014): e111265, doi:10.1371/journal.pone.0111265.

34. Elizabet saes da Silva et al., "Use of Sunscreen and Risk of Melanoma and Nonmelanoma Skin Cancer: A Systematic Review and Meta-Analysis," *European Journal of Dermatology* 28, no. 2 (2018): 186–201; Leslie K. Dennis, Laura E. Beane Freeman, and Marta J. VanBeek, "Sunscreen Use and the Risk for Melanoma: A Quantitative Review," *Annals of Internal Medicine* 139, no. 12 (2003): 966–78; Michael Huncharek and Bruce

Kupelnick, "Use of Topical Sunscreens and the Risk of Malignant Melanoma: A Meta-Analysis of 9067 Patients from 11 Case–Control Studies," *American Journal of Public Health* 92, no. 7 (2002): 1173–77.

35. J. MacLaughlin and M. F. Holick, "Aging Decreases the Capacity of Human Skin to Produce Vitamin D_3," *Journal of Clinical Investigation* 76, no. 4 (1985): 1536–38, doi: 10.1172/JCI112134.

36. J. Christopher Gallagher, "Vitamin D and Aging," *Endocrinology and Metabolism Clinics of North America* 42, no. 2 (2013): 319–32, doi:10.1016/j.ecl.2013.02.004.

37. Fahad Alshahrani and Naji Aljohani, "Vitamin D: Deficiency, Sufficiency and Toxicity," *Nutrients* 5, no. 9 (2013): 3605–16, doi:10.3390/nu5093605.

38. Emma Childs and Harriet de Wit, "Regular Exercise Is Associated with Emotional Resilience to Acute Stress in Healthy Adults," *Frontiers in Physiology* 5, no. 161 (2014), doi:10.3389/fphys.2014.00161.

39. Bruce S. McEwen and John C. Wingfield, "The Concept of Allostasis in Biology and Biomedicine," *Hormones and Behavior* 43.1 (2003): 2–15.

40. Michael T. Heneka, "Locus Ceruleus Controls Alzheimer's Disease Pathology by Modulating Microglial Functions Through Norepinephrine," *Proceedings of the National Academy of Sciences of the United States of America* 107.13 (2010): 6058–63, doi:10.1073/pnas.0909586107.

41. Joanna Rymaszewska et al., "Whole-Body Cryotherapy as Adjunct Treatment of Depressive and Anxiety Disorders," *Archivum Immunologiae et Therapiae Experimentalis* 56.1 (2008): 63–68, doi:10.1007/s00005-008-0006-5.

42. Christoffer van Tulleken et al., "Open Water Swimming as a Treatment for Major Depressive Disorder," *BMJ Case Reports* 2018 (2018), doi:10.1136/bcr-2018-225007.

43. P. Šrámek et al., "Human Physiological Responses to Immersion into Water of Different Temperatures," *European Journal of Applied Physiology* 81, no. 5 (2000): 436–42.

44. Wouter van Marken Lichtenbelt and Patrick Schrauwen, "Implications of Nonshivering Thermogenesis for Energy Balance Regulation in Humans," *American Journal of Physiology–Regulatory, Integrative and Comparative Physiology* 301, no. 2 (2011): R285–96.

45. P. Šrámek et al., "Human Physiological Responses to Immersion into Water of Different Temperatures," *European Journal of Applied Physiology* 81, no. 5 (2000): 436–42.

46. Wouter van Marken Lichtenbelt et al., "Healthy Excursions Outside the Thermal Comfort Zone," *Building Research & Information* 45, no. 7 (2017): 819–27; Mark J.W. Hanssen et al., "Short-Term Cold Acclimation Improves Insulin Sensitivity in Patients with Type 2 Diabetes Mellitus," *Nature Medicine* 21, no. 8 (2015): 863.

47. Gregory N. Bratman et al., "Nature Experience Reduces Rumination and Subgenual Prefrontal Cortex Activation," *Proceedings of the National Academy of Sciences* 112, no. 28 (2015): 8567–72.

48. Tatsuo Watanabe et al., "Green Odor and Depressive-Like State in Rats: Toward an Evidence-Based Alternative Medicine?" *Behavioural Brain Research* 224, no. 2 (2011): 290–96.

49. MaryCarol Rossiter Hunter, "Urban Nature Experiences Reduce Stress in the Context of Daily Life Based on Salivary Biomarkers," *Frontiers in Psychology* 10 (2019): 722.

50. Pascal Imbeault, Isabelle Dépault, and François Haman, "Cold Exposure Increases Adiponectin Levels in Men," *Metabolism* 58. no. 4 (2009): 552–59.

51. Arnav Katira and Peng H. Tan, "Evolving Role of Adiponectin in Cancer-Controversies and Update," *Cancer Biology & Medicine* 13, no. 1 (2016): 101.

52. Juhyun Song and Jong Eun Lee, "Adiponectin as a New Paradigm for Approaching Alzheimer's Disease," *Anatomy & Cell Biology* 46, no. 4 (2013): 229–34, doi:10.5115/acb .2013.46.4.229.

53. Tanjaniina Laukkanen et al., "Sauna Bathing Is Inversely Associated with Dementia and Alzheimer's Disease in Middle-Aged Finnish Men," *Age and Ageing* 46, no. 2 (2016): 245–49.

54. Vienna E. Brunt et al., "Passive Heat Therapy Improves Endothelial Function, Arterial Stiffness and Blood Pressure in Sedentary Humans," *Journal of Physiology* 594, no. 18 (2016): 5329–42.

55. Joy Hussain and Marc Cohen, "Clinical Effects of Regular Dry Sauna Bathing: A Systematic Review," *Evidence-Based Complementary and Alternative Medicine* 2018: 1857413, doi:10.1155/2018/1857413.

56. Małgorzata Żychowska et al., "Effects of Sauna Bathing on Stress-Related Genes Expression in Athletes and Non-athletes," *Annals of Agricultural and Environmental Medicine* 24, no. 1 (2017): 104–7.

57. Minoru Narita et al., "Heterologous μ-opioid Receptor Adaptation by Repeated Stimulation of κ-opioid Receptor: Up-regulation of G-protein Activation and Antinociception," *Journal of Neurochemistry* 85, no. 5 (2003): 1171–79.

58. Barbara A. Maher et al., "Magnetite Pollution Nanoparticles in the Human Brain," *Proceedings of the National Academy of Sciences* 113, no. 39 (2016): 10797–801.

59. Xin Zhang, Xi Chen, and Xiaobo Zhang, "The Impact of Exposure to Air Pollution on Cognitive Performance," *Proceedings of the National Academy of Sciences* 115, no. 37 (2018): 9193–97.

60. Mafalda Cacciottolo et al., "Particulate Air Pollutants, APOE Alleles and Their Contributions to Cognitive Impairment in Older Women and to Amyloidogenesis in Experimental Models," *Translational Psychiatry* 7, no. 1 (2017): e1022.

61. Jia Zhong et al., "B-vitamin Supplementation Mitigates Effects of Fine Particles on Cardiac Autonomic Dysfunction and Inflammation: A Pilot Human Intervention Trial," *Scientific Reports* 7 (2017): 45322.

62. Xiang-Yong Li et al., "Protection Against Fine Particle-Induced Pulmonary and Systemic Inflammation by Omega-3 Polyunsaturated Fatty Acids," *Biochimica et Biophysica Acta (BBA)—General Subjects* 1861, no. 3 (2017): 577–84.

63. Isabelle Romieu et al., "The Effect of Supplementation with Omega-3 Polyunsaturated Fatty Acids on Markers of Oxidative Stress in Elderly Exposed to PM(2.5)," *Environmental Health Perspectives* 116, no. 9 (2008): 1237–42.

64. David Heber et al., "Sulforaphane-Rich Broccoli Sprout Extract Attenuates Nasal Allergic Response to Diesel Exhaust Particles," *Food & Function* 5, no. 1 (2014): 35–41.

65. Patricia A. Egner et al., "Rapid and Sustainable Detoxication of Airborne Pollutants by Broccoli Sprout Beverage: Results of a Randomized Clinical Trial in China," *Cancer Prevention Research* 7, no. 8 (2014): 813–23, doi:10.1158/1940–6207.CAPR-14–0103.

66. Fabricio Pagani Possamai et al., "Antioxidant Intervention Compensates Oxidative Stress in Blood of Subjects Exposed to Emissions from a Coal Electric-Power Plant in South Brazil," *Environmental Toxicology and Pharmacology* 30, no. 2 (2010): 175–80.

4: GET OFF YO' A**

1. Steven F. Lewis and Charles H. Hennekens, "Regular Physical Activity: Forgotten Benefits," *American Journal of Medicine* 129, no. 2 (2016): 137–38.

2. Christian von Loeffelholz and Andreas Birkenfeld, "The Role of Non-exercise Activity Thermogenesis in Human Obesity," *Endotext [Internet]*, MDText.com, Inc., 2018.

3. Theodore B. Vanltallie, "Resistance to Weight Gain During Overfeeding: A NEAT Explanation," *Nutrition Reviews* 59, no. 2 (2001): 48–51.

4. James A. Levine, Norman L. Eberhardt, and Michael D. Jensen, "Role of Nonexercise Activity Thermogenesis in Resistance to Fat Gain in Humans," *Science* 283, no. 5399 (1999): 212–14.

5. Lionel Bey and Marc T. Hamilton, "Suppression of Skeletal Muscle Lipoprotein Lipase Activity During Physical Inactivity: A Molecular Reason to Maintain Daily Low-Intensity Activity," *Journal of Physiology* 551.Pt.2 (2003): 673–82, doi:10.1113/jphysiol .2003.045591.

6. M. R. Taskinen and E. A. Nikkilä, "Effect of Acute Vigorous Exercise on Lipoprotein Lipase Activity of Adipose Tissue and Skeletal Muscle in Physically Active Men," *Artery* 6, no. 6 (1980): 471–83.

7. Sophie E. Carter, et al., "Regular Walking Breaks Prevent the Decline in Cerebral Blood Flow Associated with Prolonged Sitting," *Journal of Applied Physiology* 125.3 (2018): 790–98.

8. Ira J. Goldberg et al., "Regulation of Fatty Acid Uptake into Tissues: Lipoprotein Lipase- and CD36-Mediated Pathways," *Journal of Lipid Research* 50 Suppl. (2009): S86–90, doi:10.1194/jlr.R800085-JLR200.

9. Justin R. Trombold et al., "Acute High-Intensity Endurance Exercise Is More Effective Than Moderate-Intensity Exercise for Attenuation of Postprandial Triglyceride Elevation," *Journal of Applied Physiology* 114, no. 6 (2013): 792–800.

10. Francesco Zurlo et al., "Low Ratio of Fat to Carbohydrate Oxidation as Predictor of Weight Gain: Study of 24-h RQ," *American Journal of Physiology-Endocrinology and Metabolism* 259, no. 5 (1990): E650–57.

11. Joana Araújo, Jianwen Cai, and June Stevens, "Prevalence of Optimal Metabolic Health in American Adults: National Health and Nutrition Examination Survey 2009–2016," *Metabolic Syndrome and Related Disorders* 17.1 (2019): 46–52.

12. Gian Paolo Fadini et al., "At the Crossroads of Longevity and Metabolism: The Metabolic Syndrome and Lifespan Determinant Pathways," *Aging Cell* 10, no. 1 (2011): 10–17.

13. Hidetaka Hamasaki et al., "Daily Physical Activity Assessed by a Triaxial Accelerometer Is Beneficially Associated with Waist Circumference, Serum Triglycerides, and Insulin Resistance in Japanese Patients with Prediabetes or Untreated Early Type 2 Diabetes," *Journal of Diabetes Research* 2015 (2015).

14. Elin Ekblom-Bak et al., "The Importance of Non-exercise Physical Activity for Cardiovascular Health and Longevity," *British Journal of Sports Medicine* 48, no. 3 (2014): 233–38.

15. Bernard M. F. M. Duvivier et al., "Minimal Intensity Physical Activity (Standing and Walking) of Longer Duration Improves Insulin Action and Plasma Lipids More Than Shorter Periods of Moderate to Vigorous Exercise (Cycling) in Sedentary Subjects When Energy Expenditure Is Comparable," *PLOS ONE* 8, no. 2 (2013): e55542.

16. Carter et al., "Regular Walking Breaks."

17. Ernest R. Greene, Kushum Shrestha, and Analyssa Garcia, "Acute Effects of Walking on Human Internal Carotid Blood Flow," *FASEB Journal* 31, no. 1 Suppl. (2017): 840–23.

18. Chun Liang Hsu et al., "Aerobic Exercise Promotes Executive Functions and Impacts Functional Neural Activity Among Older Adults with Vascular Cognitive Impairment," *British Journal of Sports Medicine* 52, no. 3 (2018): 184–91.

19. Aron S. Buchman et al., "Physical Activity, Common Brain Pathologies, and Cognition in Community-Dwelling Older Adults," *Neurology* 92, no. 8 (2019): e811–22.

20. Mark A. Hearris et al., "Regulation of Muscle Glycogen Metabolism During Exercise: Implications for Endurance Performance and Training Adaptations," *Nutrients* 10, no. 3 (2018): 298, doi:10.3390/nu10030298.

21. Brad Jon Schoenfeld and Alan Albert Aragon, "How Much Protein Can the Body Use in a Single Meal for Muscle-Building? Implications for Daily Protein Distribution," *Journal of the International Society of Sports Nutrition* 15, no. 1 (2018): 10.

22. Alan Albert Aragon and Brad Jon Schoenfeld, "Nutrient Timing Revisited: Is There a Post-exercise Anabolic Window?" *Journal of the International Society of Sports Nutrition* 10, no. 1 (2013): 5.

23. Ibid.

24. George A. Brooks, "Cell–Cell and Intracellular Lactate Shuttles," *Journal of Physiology* 587.Pt.23 (2009): 5591–600, doi:10.1113/jphysiol.2009.178350.

25. Patrizia Proia et al., "Lactate as a Metabolite and a Regulator in the Central Nervous System," *International Journal of Molecular Sciences* 17, no. 9 (2016): 1450, doi:10.3390/ijms17091450.

26. Laurel Riske et al., "Lactate in the Brain: An Update on Its Relevance to Brain Energy, Neurons, Glia and Panic Disorder," *Therapeutic Advances in Psychopharmacology* 7, no. 2 (2016): 85–89, doi:10.1177/2045125316675579.

27. Proia et al., "Lactate as a Metabolite."

28. Margaret Schenkman et al., "Effect of High-Intensity Treadmill Exercise on Motor Symptoms in Patients with de Novo Parkinson's Disease: A Phase 2 Randomized Clinical Trial," *JAMA Neurology* 75, no. 2 (2018): 219–26.

29. Jenna B. Gillen et al., "Twelve Weeks of Sprint Interval Training Improves Indices of Cardiometabolic Health Similar to Traditional Endurance Training Despite a Five-fold Lower Exercise Volume and Time Commitment," *PLOS ONE* 11, no. 4 (2016): e0154075.

30. Robert Acton Jacobs et al., "Improvements in Exercise Performance with High-Intensity Interval Training Coincide with an Increase in Skeletal Muscle Mitochondrial Content and Function," *Journal of Applied Physiology* 115, no. 6 (2013): 785–93.

31. Masahiro Banno et al., "Exercise Can Improve Sleep Quality: A Systematic Review and Meta-Analysis," *PeerJ* 6 (2018): e5172, doi: 10.7717/peerj.5172.

32. Joseph T. Flynn et al., "Clinical Practice Guideline for Screening and Management of High Blood Pressure in Children and Adolescents," *Pediatrics* 140, no. 3 (2017): e20171904.

33. Jeff D. Williamson et al., "Effect of Intensive vs. Standard Blood Pressure Control on Probable Dementia: A Randomized Clinical Trial," *JAMA* 321, no. 6 (2019): 553–61.

34. Lisa A. Te Morenga et al., "Dietary Sugars and Cardiometabolic Risk: Systematic Review and Meta-Analyses of Randomized Controlled Trials of the Effects on Blood Pressure and Lipids," *American Journal of Clinical Nutrition* 100, no. 1 (2014): 65–79.

35. Tessio Rebello, Robert E. Hodges, and Jack L. Smith, "Short-Term Effects of Various Sugars on Antinatriuresis and Blood Pressure Changes in Normotensive Young Men," *American Journal of Clinical Nutrition* 38, no. 1 (1983): 84–94.

36. Huseyin Naci et al., "How Does Exercise Treatment Compare with Antihypertensive Medications? A Network Meta-Analysis of 391 Randomised Controlled Trials Assessing Exercise and Medication Effects on Systolic Blood Pressure," *British Journal of Sports Medicine* (2018): 53, (2018): 859–69.

37. Eric D. Vidoni et al., "Dose-Response of Aerobic Exercise on Cognition: A Community-Based, Pilot Randomized Controlled Trial," *PLOS ONE* 10, no. 7 (2015): e0131647, doi:10.1371/journal.pone.0131647.

38. Lin Li et al., "Acute Aerobic Exercise Increases Cortical Activity During Working Memory: A Functional MRI Study in Female College Students," *PLOS ONE* 9, no. 6 (2014): e99222.

39. Fengqin Liu et al., "It Takes Biking to Learn: Physical Activity Improves Learning a Second Language," *PLOS ONE* 12, no. 5 (2017): e0177624.

40. Felipe B. Schuch et al., "Are Lower Levels of Cardiorespiratory Fitness Associated with Incident Depression? A Systematic Review of Prospective Cohort Studies," *Preventive Medicine* 93 (2016): 159–65.

41. Ioannis D. Morres et al., "Aerobic Exercise for Adult Patients with Major Depressive Disorder in Mental Health Services: A Systematic Review and Meta-Analysis," *Depression and Anxiety* 36, no. 1 (2019): 39–53.

42. Brett R. Gordon et al., "Association of Efficacy of Resistance Exercise Training with Depressive Symptoms: Meta-Analysis and Meta-Regression Analysis of Randomized Clinical Trials," *JAMA Psychiatry* 75, no. 6 (2018): 566–76.

43. Brett R. Gordon et al., "The Effects of Resistance Exercise Training on Anxiety: A Meta-Analysis and Meta-Regression Analysis of Randomized Controlled Trials," *Sports Medicine* 47, no. 12 (2017): 2521–32.

44. Friederike Klempin et al., "Serotonin Is Required for Exercise-Induced Adult Hippocampal Neurogenesis," *Journal of Neuroscience* 33, no. 19 (2013): 8270–75.

45. Kristen M. Beavers et al., "Effect of Exercise Type During Intentional Weight Loss on Body Composition in Older Adults with Obesity," *Obesity* 25, no. 11 (2017): 1823–29, doi:10.1002/oby.21977.

46. Emmanuel Stamatakis et al., "Does Strength-Promoting Exercise Confer Unique Health Benefits? A Pooled Analysis of Data on 11 Population Cohorts with All-Cause, Cancer, and Cardiovascular Mortality Endpoints," *American Journal of Epidemiology* 187, no. 5 (2017): 1102–12.

47. Yorgi Mavros et al., "Mediation of Cognitive Function Improvements by Strength Gains After Resistance Training in Older Adults with Mild Cognitive Impairment: Outcomes of the Study of Mental and Resistance Training," *Journal of the American Geriatrics Society* 65, no. 3 (2017): 550–59.

48. Ivan Bautmans, Katrien Van Puyvelde, and Tony Mets, "Sarcopenia and Functional Decline: Pathophysiology, Prevention and Therapy" *Acta Clinica Belgica* 64, no. 4 (2009): 303–16.

49. Monique E. Francois et al., "'Exercise Snacks' Before Meals: A Novel Strategy to Improve Glycaemic Control in Individuals with Insulin Resistance," *Diabetologia* 57, no. 7 (2014): 1437–45.

50. Brad J. Schoenfeld et al., "Influence of Resistance Training Frequency on Muscular Adaptations in Well-Trained Men," *Journal of Strength & Conditioning Research* 29, no. 7 (2015): 1821–29.

51. Laura D. Baker et al., "Effects of Growth Hormone–Releasing Hormone on Cognitive

Function in Adults with Mild Cognitive Impairment and Healthy Older Adults: Results of a Controlled Trial," *Archives of Neurology* 69, no. 11 (2012): 1420–29, doi:10.1001/archneurol.2012.1970.

52. Gabrielle Brandenberger et al., "Effect of Sleep Deprivation on Overall 24 h Growth-Hormone Secretion," *Lancet* 356, no. 9239 (2000): 1408.

53. Johanna A. Pallotta, and Patricia J. Kennedy, "Response of Plasma Insulin and Growth Hormone to Carbohydrate and Protein Feeding," *Metabolism* 17.10 (1968): 901–8.

54. Helene Nørrelund, "The Metabolic Role of Growth Hormone in Humans with Particular Reference to Fasting," *Growth Hormone & IGF Research* 15, no. 2 (2005): 95–122.

55. Rachel Leproult and Eve Van Cauter, "Effect of 1 Week of Sleep Restriction on Testosterone Levels in Young Healthy Men," *JAMA* 305, no. 21 (2011): 2173–4, doi:10.1001/jama.2011.710.

56. Flavio A. Cadegiani and Claudio E. Kater, "Hormonal Aspects of Overtraining Syndrome: A Systematic Review," *BMC Sports Science, Medicine & Rehabilitation* 9, no. 14 (2017), doi:10.1186/s13102-017-0079-8.

57. Nathaniel D. M. Jenkins et al., "Greater Neural Adaptations Following High- vs. Low-Load Resistance Training," *Frontiers in Physiology* 8 (2017): 331.

5: TOXIC WORLD

1. Robert Dales et al., "Quality of Indoor Residential Air and Health," *CMAJ: Canadian Medical Association Journal* 179, no. 2 (2008): 147–52, doi:10.1503/cmaj.070359.

2. "Bisphenol A (BPA)," *National Institute of Environmental Health Sciences*, U.S. Department of Health and Human Services, www.niehs.nih.gov/health/topics/agents/sya-bpa/index.cfm; Buyun Liu et al., "Bisphenol A Substitutes and Obesity in US Adults: Analysis of a Population-Based, Cross-Sectional Study," *Lancet, Planetary Health* 1.3 (2017): e114–22, doi:10.1016/S2542-5196(17)30049-9.

3. Rachael Beairsto, "Is BPA Safe? Endocrine Society Addresses FDA Position on Commercial BPA Use," *Endocrinology Advisor*, October 24, 2018, www.endocrinologyadvisor.com/home/topics/general-endocrinology/is-bpa-safe-endocrine-society-addresses-fda-position-on-commercial-bpa-use/.

4. Subhrangsu S. Mandal, ed., *Gene Regulation, Epigenetics and Hormone Signaling*, vol. 1 (TK: John Wiley & Sons, 2017).

5. Julia R. Varshavsky et al., "Dietary Sources of Cumulative Phthalates Exposure Among the US General Population in NHANES 2005–2014," *Environment International* 115 (2018): 417–29.

6. Kristen M. Rappazzo et al., "Exposure to Perfluorinated Alkyl Substances and Health Outcomes in Children: A Systematic Review of the Epidemiologic Literature," *International Journal of Environmental Research and Public Health* 14, no. 7 (2017): 691, doi:10.3390/ijerph14070691; Gang Liu et al., "Perfluoroalkyl Substances and Changes in Body Weight and Resting Metabolic Rate in Response to Weight-Loss Diets: A Prospective Study," *PLOS Medicine* 15, no. 2 (2018): e1002502, doi:10.1371/journal.pmed.1002502.

7. Ying Li et al., "Half-Lives of PFOS, PFHxS and PFOA After End of Exposure to Contaminated Drinking Water," *Occupational and Environmental Medicine* 75, no. 1 (2018): 46–51.

8. Katherine E. Boronow et al., "Serum Concentrations of PFASs and Exposure-Related Behaviors in African American and Non-Hispanic White Women," *Journal of Exposure Science & Environmental Epidemiology* 29, no. 2 (2019): 206.

9. Patricia Callahan and Sam Roe, "Big Tobacco Wins Fire Marshals as Allies in Flame Retardant Push," chicagotribune.com, March 21, 2019, www.chicagotribune.com/ct -met-flames-tobacco-20120508-story.html.

10. Julie B. Herbstman et al., "Prenatal Exposure to PBDEs and Neurodevelopment," *Environmental Health Perspectives* 118, no. 5 (2010): 712–19.

11. Carla A. Ng et al., "Polybrominated Diphenyl Ether (PBDE) Accumulation in Farmed Salmon Evaluated Using a Dynamic Sea-Cage Production Model," *Environmental Science & Technology* 52, no. 12 (2018): 6965–73.

12. Sumedha M. Joshi, "The Sick Building Syndrome," *Indian Journal of Occupational and Environmental Medicine* 12, no. 2 (2008): 61–64, doi:10.4103/0019–5278.43262.

13. P. D. Darbre et al., "Concentrations of Parabens in Human Breast Tumours," *Journal of Applied Toxicology* 24, no. 1 (2004): 5–13.

14. Damian Maseda et al., "Nonsteroidal Anti-inflammatory Drugs Alter the Microbiota and Exacerbate *Clostridium difficile* Colitis While Dysregulating the Inflammatory Response," *mBio* 10, no. 1 (2019): e02282–18, doi:10.1128/mBio.02282–18.

15. Mats Lilja et al., "High Doses of Anti-inflammatory Drugs Compromise Muscle Strength and Hypertrophic Adaptations to Resistance Training in Young Adults," *Acta Physiologica* 222, no. 2 (2018): e12948.

16. Dominik Mischkowski, Jennifer Crocker, and Baldwin M. Way, "From Painkiller to Empathy Killer: Acetaminophen (Paracetamol) Reduces Empathy for Pain," *Social Cognitive and Affective Neuroscience* 11, no. 9 (2016): 1345–53.

17. Claudia B. Avella-Garcia et al., "Acetaminophen Use in Pregnancy and Neurodevelopment: Attention Function and Autism Spectrum Symptoms," *International Journal of Epidemiology* 45, no. 6 (2016): 1987–96.

18. C. G. Bornehag et al., "Prenatal Exposure to Acetaminophen and Children's Language Development at 30 Months," *European Psychiatry* 51 (2018): 98–103.

19. John T. Slattery et al., "Dose-Dependent Pharmacokinetics of Acetaminophen: Evidence of Glutathione Depletion in Humans," *Clinical Pharmacology & Therapeutics* 41, no. 4 (1987): 413–18.

20. Xueya Cai et al., "Long-Term Anticholinergic Use and the Aging Brain," *Alzheimer's & Dementia* 9, no. 4 (2013): 377–85, doi:10.1016/j.jalz.2012.02.005.

21. Shelly L. Gray et al., "Cumulative Use of Strong Anticholinergics and Incident Dementia: A Prospective Cohort Study," *JAMA Internal Medicine* 175, no. 3 (2015): 401–7, doi:10.1001/jamainternmed.2014.7663.

22. Ghada Bassioni et al., "Risk Assessment of Using Aluminum Foil in Food Preparation," *International Journal of Electrochemical Science* 7, no. 5 (2012): 4498–509.

23. Clare Minshall, Jodie Nadal, and Christopher Exley, "Aluminium in Human Sweat," *Journal of Trace Elements in Medicine and Biology* 28, no. 1 (2014): 87–88.

24. Pranita D. Tamma and Sara E. Cosgrove, "Addressing the Appropriateness of Outpatient Antibiotic Prescribing in the United States: An Important First Step," *JAMA* 315, no. 17 (2016): 1839–41.

25. Jordan E. Bisanz et al., "Randomized Open-Label Pilot Study of the Influence of Probiotics and the Gut Microbiome on Toxic Metal Levels in Tanzanian Pregnant Women and School Children," *mBio* 5, no. 5 (2014): e01580–14.

26. Les Dethlefsen et al., "The Pervasive Effects of an Antibiotic on the Human Gut Microbiota, as Revealed by Deep 16S rRNA Sequencing," *PLOS Biology* 6, no. 11 (2008): e280, doi:10.1371/journal.pbio.0060280.

27. Tsepo Ramatla et al., "Evaluation of Antibiotic Residues in Raw Meat Using Different Analytical Methods," *Antibiotics* 6.4 (2017): 34, doi:10.3390/antibiotics6040034; Khurram Muaz et al., "Antibiotic Residues in Chicken Meat: Global Prevalence, Threats, and Decontamination Strategies: A Review," *Journal of Food Protection* 81, no. 4 (2018): 619–27.

28. Marcin Barański et al., "Higher Antioxidant and Lower Cadmium Concentrations and Lower Incidence of Pesticide Residues in Organically Grown Crops: A Systematic Literature Review and Meta-Analyses," *British Journal of Nutrition* 112, no. 5 (2014): 794–811.

29. Jotham Suez et al., "Post-antibiotic Gut Mucosal Microbiome Reconstitution Is Impaired by Probiotics and Improved by Autologous FMT," *Cell* 174, no. 6 (2018): 1406–23.

30. Ruth E. Brown et al., "Secular Differences in the Association Between Caloric Intake, Macronutrient Intake, and Physical Activity with Obesity," *Obesity Research & Clinical Practice* 10, no. 3 (2016): 243–55.

31. Tetsuhide Ito and Robert T. Jensen, "Association of Long-Term Proton Pump Inhibitor Therapy with Bone Fractures and Effects on Absorption of Calcium, Vitamin B_{12}, Iron, and Magnesium," *Current Gastroenterology Reports* 12, no. 6 (2010): 448–57, doi:10.1007/s11894-010-0141-0.

32. Elizabet saes da Silva et al., "Use of Sunscreen and Risk of Melanoma and Nonmelanoma Skin Cancer: A Systematic Review and Meta-Analysis," *European Journal of Dermatology* 28, no. 2 (2018): 186–201; Leslie K. Dennis, Laura E. Beane Freeman, and Marta J. VanBeek, "Sunscreen Use and the Risk for Melanoma: A Quantitative Review," *Annals of Internal Medicine* 139, no. 12 (2003): 966–78; Michael Huncharek and Bruce Kupelnick, "Use of Topical Sunscreens and the Risk of Malignant Melanoma: A Meta-Analysis of 9067 Patients from 11 Case-Control Studies," *American Journal of Public Health* 92, no. 7 (2002): 1173–77.

33. Cheng Wang et al., "Stability and Removal of Selected Avobenzone's Chlorination Products," *Chemosphere* 182 (2017): 238–44.

34. Murali K. Matta et al., "Effect of Sunscreen Application Under Maximal Use Conditions on Plasma Concentration of Sunscreen Active Ingredients: A Randomized Clinical Trial," *JAMA* 321, no. 21 (2019): 2082–91.

35. Naoki Ito et al., "The Protective Role of Astaxanthin for UV-Induced Skin Deterioration in Healthy People—A Randomized, Double-Blind, Placebo-Controlled Trial," *Nutrients* 10, no. 7 (2018): 817, doi:10.3390/nu10070817.

36. Rui Li et al., "Mercury Pollution in Vegetables, Grains and Soils from Areas Surrounding Coal-Fired Power Plants," *Scientific Reports* 7, no. 46545 (2017).

37. Nicholas V. C. Ralston et al., "Dietary Selenium's Protective Effects Against Methylmercury Toxicity." *Toxicology* 278.1 (2010): 112–123.

38. Philippe Grandjean et al., "Cognitive Deficit in 7-Year-Old Children with Prenatal Exposure to Methylmercury," *Neurotoxicology and Teratology* 19, no. 6 (1997): 417–28.

39. Ondine van de Rest et al., "APOE ε4 and the Associations of Seafood and Long-Chain Omega-3 Fatty Acids with Cognitive Decline," *Neurology* 86, no. 22 (2016): 2063–70.

40. Martha Clare Morris et al., "Association of Seafood Consumption, Brain Mercury Level, and APOE ε4 Status with Brain Neuropathology in Older Adults," *JAMA* 315, no. 5 (2016): 489–97, doi:10.1001/jama.2015.19451.

41. Jianghong Liu et al., "The Mediating Role of Sleep in the Fish Consumption–Cognitive Functioning Relationship: A Cohort Study," *Scientific Reports* 7, no. 1 (2017): 17961;

Joseph R. Hibbeln et al. "Maternal Seafood Consumption in Pregnancy and Neurodevelopmental Outcomes in Childhood (ALSPAC Study): An Observational Cohort Study." *The Lancet* 369.9561 (2007): 578–585.

42. Maria A. I. Åberg et al., "Fish Intake of Swedish Male Adolescents Is a Predictor of Cognitive Performance," *Acta Paediatrica* 98, no. 3 (2009): 555–60.

43. Margaret E. Sears et al., "Arsenic, Cadmium, Lead, and Mercury in Sweat: A Systematic Review," *Journal of Environmental and Public Health* 2012, no. 184745 (2012), doi:10.1155/2012/184745.

44. T. T. Sjursen et al., "Changes in Health Complaints After Removal of Amalgam Fillings," *Journal of Oral Rehabilitation* 38, no. 11 (2011): 835–48, doi:10.1111/j.1365–2842 .2011.02223.x.

45. R. C. Kaltreider et al., "Arsenic Alters the Function of the Glucocorticoid Receptor as a Transcription Factor," *Environmental Health Perspectives* 109, no. 3 (2001): 245–51, doi:10.1289/ehp.01109245

46. Frederick M. Fishel, *Pesticide Use Trends in the United States: Agricultural Pesticides,* University of Florida IFSAS Extension, http://edis.ifas.ufl.edu/pi176.

47. Isioma Tongo and Lawrence Ezemonye, "Human Health Risks Associated with Residual Pesticide Levels in Edible Tissues of Slaughtered Cattle in Benin City, Southern Nigeria," *Toxicology Reports* 3, no. 2 (2015): 1117–35, doi:10.1016/j.toxrep.2015.07.008.

48. Wissem Mnif et al., "Effect of Endocrine Disruptor Pesticides: A Review," *International Journal of Environmental Research and Public Health* 8, no. 6 (2011): 2265–303, doi:10.3390/ijerph8062265.

49. Carly Hyland et al., "Organic Diet Intervention Significantly Reduces Urinary Pesticide Levels in U.S. Children and Adults," *Environmental Research* 171 (2019): 568–75.

50. Julia Baudry et al., "Association of Frequency of Organic Food Consumption with Cancer Risk: Findings from the NutriNet-Santé Prospective Cohort Study," *JAMA Internal Medicine* 178, no. 12 (2018): 1597–606; Luoping Zhang et al., "Exposure to Glyphosate-Based Herbicides and Risk for Non-Hodgkin Lymphoma: A Meta-analysis and Supporting Evidence," *Mutation Research/Reviews in Mutation Research* (2019).

51. Timothy Ciesielski et al., "Cadmium Exposure and Neurodevelopmental Outcomes in U.S. Children," *Environmental Health Perspectives* 120, no. 5 (2012): 758–63, doi:10.1289 /ehp.1104152.

52. Marcin Barański et al., "Higher Antioxidant and Lower Cadmium Concentrations and Lower Incidence of Pesticide Residues in Organically Grown Crops: A Systematic Literature Review and Meta-analyses," *British Journal of Nutrition* 112, no. 5 (2014): 794–811.

53. Rodjana Chunhabundit, "Cadmium Exposure and Potential Health Risk from Foods in Contaminated Area, Thailand," *Toxicological Research* 32, no. 1 (2016): 65–72, doi: 10.5487/TR.2016.32.1.065.

54. "New Study Finds Lead Levels in a Majority of Paints Exceed Chinese Regulation and Should Not Be on Store Shelves," *IPEN*, ipen.org/news/new-study-finds-lead-levels -majority-paints-exceed-chinese-regulation-and-should-not-be-store.

55. "Lead in Food: A Hidden Health Threat," Environmental Defense Fund, EDF Health, June 15, 2017.

56. "Health Effects of Low-Level Lead Evaluation," National Toxicology Program, National Institute of Environmental Health Sciences, U.S. Department of Health and Human Services, ntp.niehs.nih.gov/pubhealth/hat/noms/lead/index.html.

57. Olukayode Okunade et al., "Supplementation of the Diet by Exogenous Myrosinase via Mustard Seeds to Increase the Bioavailability of Sulforaphane in Healthy Human Subjects After the Consumption of Cooked Broccoli," *Molecular Nutrition & Food Research* 62, no. 18 (2018): 69 1700980.

58. J. W. Fahey et al., "Broccoli Sprouts: An Exceptionally Rich Source of Inducers of Enzymes That Protect Against Chemical Carcinogens," *Proceedings of the National Academy of Sciences of the United States of America* 94, no. 19 (1997): 10367–72, doi:10.1073/pnas.94.19.10367.

59. Michael C. Petriello et al., "Modulation of Persistent Organic Pollutant Toxicity Through Nutritional Intervention: Emerging Opportunities in Biomedicine and Environmental Remediation," *Science of the Total Environment* 491–492 (2014): 11–16, doi:10.1016/j.scitotenv.2014.01.109.

60. K. D. Kent, W. J. Harper, and J. A. Bomser, "Effect of Whey Protein Isolate on Intracellular Glutathione and Oxidant-Induced Cell Death in Human Prostate Epithelial Cells," *Toxicology in Vitro* 17, no. 1 (2003): 27–33.

6: PEACE OF MIND

1. Michael G. Gottschalk and Katharina Domschke, "Genetics of Generalized Anxiety Disorder and Related Traits," *Dialogues in Clinical Neuroscience* 19, no. 2 (2017): 159–68; Falk W. Lohoff, "Overview of the Genetics of Major Depressive Disorder," *Current Psychiatry Reports* 12, no. 6 (2010): 539–46, doi:10.1007/s11920-010-0150-6.

2. Andrea H. Weinberger et al., "Trends in Depression Prevalence in the USA from 2005 to 2015: Widening Disparities in Vulnerable Groups," *Psychological Medicine* 48, no. 8 (2018): 1308–15.

3. Conor J. Wild et al., "Dissociable Effects of Self-Reported Daily Sleep Duration on High-Level Cognitive Abilities," *Sleep* 41, no. 12 (2018), doi:10.1093/sleep/zsy182.

4. Esther Donga et al., "A Single Night of Partial Sleep Deprivation Induces Insulin Resistance in Multiple Metabolic Pathways in Healthy Subjects," *Journal of Clinical Endocrinology & Metabolism* 95, no. 6 (2010): 2963–68.

5. Jerrah K. Holth et al., "The Sleep-Wake Cycle Regulates Brain Interstitial Fluid Tau in Mice and CSF Tau in Humans," *Science* 363, no. 6429 (2019): 880–84.

6. Thomas J. Moore and Donald R. Mattison, "Adult Utilization of Psychiatric Drugs and Differences by Sex, Age, and Race," *JAMA Internal Medicine* 177, no. 2 (2017): 274–75.

7. Seung-Schik Yoo et al., "The Human Emotional Brain without Sleep—A Prefrontal Amygdala Disconnect," *Current Biology* 17.20 (2007): R877–78.

8. Haya Al Khatib, S. V. Harding, J. Darzi, and G. K. Pot. "The Effects of Partial Sleep Deprivation on Energy Balance: A Systematic Review and Meta-Analysis." *European Journal of Clinical Nutrition* 71, no. 5 (2017): 614; Jenny Theorell-Haglöw et al., "Sleep Duration is Associated with Healthy Diet Scores and Meal Patterns: Results from the Population-Based EpiHealth Study." *Journal of Clinical Sleep Medicine* (2019).

9. Tony T. Yang et al., "Adolescents with Major Depression Demonstrate Increased Amygdala Activation," *Journal of the American Academy of Child and Adolescent Psychiatry* 49, no. 1 (2010): 42–51.

10. Yoo et al., "The Human Emotional Brain Without Sleep."

11. Eti Ben Simon and Matthew P. Walker, "Sleep Loss Causes Social Withdrawal and Loneliness," *Nature Communications* 9, no. 3146 (2018).

12. Seung-Gul Kang et al., "Decrease in fMRI Brain Activation During Working Memory Performed after Sleeping Under 10 Lux Light," *Scientific Reports* 6, no. 36731 (2016).

13. Brendan M. Gabriel and Juleen R. Zierath, "Circadian Rhythms and Exercise—Re-setting the Clock in Metabolic Disease," *Nature Reviews Endocrinology* 15, no. 4 (2019): 197–06.

14. Behnood Abbasi et al., "The Effect of Magnesium Supplementation on Primary Insomnia in Elderly: A Double-Blind Placebo-Controlled Clinical Trial," *Journal of Research in Medical Sciences: The Official Journal of Isfahan University of Medical Sciences* 17, no. 12 (2012): 1161–69.

15. Cibele Aparecida Crispim et al., "Relationship Between Food Intake and Sleep Pattern in Healthy Individuals," *Journal of Clinical Sleep Medicine* 7, no. 6 (2011): 659–64, doi:10.5664/jcsm.1476.

16. Adrian F. Ward et al., "Brain Drain: The Mere Presence of One's Own Smartphone Reduces Available Cognitive Capacity," *Journal of the Association for Consumer Research* 2, no. 2 (2017): 140–54.

17. Ji-Won Chun et al., "Role of Frontostriatal Connectivity in Adolescents with Excessive Smartphone Use," *Frontiers in Psychiatry* 9, no. 437 (2018): doi:10.3389/fpsyt.2018.00437.

18. Melissa G. Hunt et al., "No More FOMO: Limiting Social Media Decreases Loneliness and Depression," *Journal of Social and Clinical Psychology* 37, no. 10 (2018): 751–68.

19. Matteo Bergami et al., "A Critical Period for Experience-Dependent Remodeling of Adult-Born Neuron Connectivity," *Neuron* 85, no. 4 (2015): 710–17.

20. Jennifer E. Stellar et al., "Positive Affect and Markers of Inflammation: Discrete Positive Emotions Predict Lower Levels of Inflammatory Cytokines," *Emotion* 1, no. 2 (2015): 129.

21. Norman C. Reynolds Jr. and Robert Montgomery, "Using the Argonne Diet in Jet Lag Prevention: Deployment of Troops Across Nine Time Zones," *Military Medicine* 167, no. 6 (2002): 451–53.

22. Andrew Herxheimer and Keith J. Petrie, "Melatonin for the Prevention and Treatment of Jet Lag," *Cochrane Database of Systematic Reviews* 2 (2002).

23. Enzo Tagliazucchi et al., "Increased Global Functional Connectivity Correlates with LSD-Induced Ego Dissolution," *Current Biology* 26, no. 8 (2016): 1043–50.

24. Julie Scharper, "Crash Course in the Nature of Mind," *Hub*, September 1 2017, hub.jhu.edu/magazine/2017/fall/roland-griffiths-magic-mushrooms-experiment-psilocybin-depression/.

25. Tanja Miller and Laila Nielsen, "Measure of Significance of Holotropic Breathwork in the Development of Self-Awareness," *Journal of Alternative and Complementary Medicine* 21, no. 12 (2015): 796–803.

26. Mette Sørensen et al., "Long-Term Exposure to Road Traffic Noise and Incident Diabetes: A Cohort Study," *Environmental Health Perspectives* 121, no. 2 (2013): 217–22, doi:10.1289/ehp.1205503.

27. Manfred E. Beutel et al., "Noise Annoyance Is Associated with Depression and Anxiety in the General Population—The Contribution of Aircraft Noise," *PLOS ONE* 11, no. 5 (2016): e0155357, doi:10.1371/journal.pone.0155357.

28. Ivana Buric et al., "What Is the Molecular Signature of Mind-Body Interventions? A Systematic Review of Gene Expression Changes Induced by Meditation and Related Practices," *Frontiers in Immunology* 8, no. 670 (2017), doi:10.3389/fimmu.2017.00670.

29. Nicola S. Schutte and John M. Malouff, "A Meta-Analytic Review of the Effects of Mindfulness Meditation on Telomerase Activity," *Psychoneuroendocrinology* 42 (2014): 45–48.

30. Julia C. Basso et al., "Brief, Daily Meditation Enhances Attention, Memory, Mood, and Emotional Regulation in Non-experienced Meditators," *Behavioural Brain Research* 356 (2019): 208–20.

31. Ibid.

32. David R. Kille, Amanda L. Forest, and Joanne V. Wood, "Tall, Dark, and Stable: Embodiment Motivates Mate Selection Preferences," *Psychological Science* 24, no. 1 (2013): 112–14.

33. Cigna, "Cigna's U.S. Lonliness Index," https://www.multivu.com/players/English/8294451-cigna-us-loneliness-survey/.

34. American Psychological Association, "So Lonely I Could Die," August 5, 2017, https://www.apa.org/news/press/releases/2017/08/lonely-die.

7: PUTTING IT ALL TOGETHER

1. Miao-Chuan Chen, Shu-Hui Fang, and Li Fang, "The Effects of Aromatherapy in Relieving Symptoms Related to Job Stress Among Nurses," *International Journal of Nursing Practice* 21, no. 1 (2015): 87–93.

2. Alessio Fasano, "Zonulin and Its Regulation of Intestinal Barrier Function: The Biological Door to Inflammation, Autoimmunity, and Cancer," *Physiological Reviews* 91, no. 1 (2011): 151–75.

3. Gitanjali M. Singh et al., "Estimated Global, Regional, and National Disease Burdens Related to Sugar-Sweetened Beverage Consumption in 2010," *Circulation* 132, no. 8 (2015): 639–66.

4. Anallely López-Yerena et al., "Effects of Organic and Conventional Growing Systems on the Phenolic Profile of Extra-Virgin Olive Oil," *Molecules* 24, no. 10 (2019): 1986.

5. Michele Drehmer et al., "Total and Full-Fat, but Not Low-Fat, Dairy Product Intakes Are Inversely Associated with Metabolic Syndrome in Adults," *Journal of Nutrition* 146, no. 1 (2015): 81–89.

6. Alpana P. Shukla et al., "Effect of Food Order on Ghrelin Suppression," *Diabetes Care* 41, no. 5 (2018): e76–77, doi:10.2337/dc17-2244.

7. Nathalie Pross, "Effects of Dehydration on Brain Functioning: A Life-span Perspective," *Annals of Nutrition and Metabolism* 70 Suppl.1 (2017): 30–36.

8. Song Yi Park et al., "Association of Coffee Consumption with Total and Cause-Specific Mortality Among Nonwhite Populations," *Annals of Internal Medicine* 167, no. 4 (2017): 228–35.

9. Elizabeth A. Thomas et al., "Usual Breakfast Eating Habits Affect Response to Breakfast Skipping in Overweight Women," *Obesity* 23, no. 4 (2015): 750–59, doi:10.1002/oby.21049.

10. Elizabeth F. Sutton et al., "Early Time-Restricted Feeding Improves Insulin Sensitivity, Blood Pressure, and Oxidative Stress Even Without Weight Loss in Men with Prediabetes," *Cell Metabolism* 27, no. 6 (2018): 1212–21, doi:10.1016/j.cmet.2018.04.010.

11. Renata Fiche da Mata Gonçalves, et al., "Smartphone Use While Eating Increases Caloric Ingestion," *Physiology & Behavior* 204 (2019): 93–99.

Join the Cortex, a Facebook Community for Geniuses

http://maxl.ug/thecortex

Need clarity? Or just want to connect? The first place you should go is The Cortex. I've created a private Facebook community for those going through their own health journeys to share tips, tricks, recipes, research, and more. Many of them are experienced and are living the Genius Life, while others are just starting out. Make sure to introduce yourself!

Watch My Documentary, *Bread Head*

www.breadheadmovie.com

My story is documented in my film, *Bread Head*, the first and only feature-length documentary solely about dementia prevention, because changes begin in the brain decades before the first symptom of memory loss. Check out the website to support the film, see a trailer, sign up for screening alerts, and become a *Bread Head* activist.

Join My Official Newsletter

www.maxlugavere.com

Want research broken down and delivered straight to your inbox? My newsletter is where I regularly share research articles (with easy-to-read summaries), impromptu interviews, and other easily digestible tidbits designed to improve your life. No spam ever, and you can opt out anytime.

Product Suggestions

http://maxl.ug/TGLresources

Want to know my favorite blue light–blocking glasses? How about the online meditation course I recommend? On the market for a new air filter or water filter? Or perhaps a better salt option? (None of these would surprise me after you've read this book.) Over the years I have become friendly with many product manufacturers, and as a result get to try out lots of products. Here you can check out my recommendations for products alluded to in this book (and get exclusive discounts). Anything I recommend is something I've vetted and personally use.

Research Resources

One of the top ways that you can ensure the information you're getting is sound is to make sure the places you're looking for it are credible and as close to the science as possible. These are the only sources I recommend using to track and search scientific research:

ScienceDaily
www.sciencedaily.com

This site republishes university press releases that often accompany study publications. It brings together research from many different disciplines, and you can often find good stuff here by scrolling down to "Health News" or clicking "Health" in the menu bar at the top.

Note: press releases from universities are not necessarily perfect, but they are a great starting place and usually provide links to the research discussed. Reading both the press release and the study paper can help you learn how to interpret research. And the releases are often the very sources that journalists will use to write their articles. So in essence, this site takes you straight to the source!

Medical Xpress
www.medicalxpress.com

This site does the same as ScienceDaily, but is exclusively medical/health-related.

EurekAlert!
www.eurekalert.org
This is similar to the previous two resources—it publishes press releases—but is run by the American Association for the Advancement of Science, which publishes the journal *Science*.

PubMed
www.ncbi.nlm.nih.gov/pubmed
When researching, I often use PubMed. One way to use Google to search PubMed is to add "site:nih.gov" into your Google search. For example, "Alzheimer's insulin site:nih.gov" would search the NIH website (which includes PubMed) for all articles mentioning Alzheimer's and insulin.

Contact Me
Contact me for speaking or coaching, or to just say hi!

www.maxlugavere.com

info@maxlugavere.com

instagram.com/maxlugavere

facebook.com/maxlugavere

twitter.com/maxlugavere

INDEX

Note: Italic page numbers refer to charts.

ABOUT THE AUTHOR

Max Lugavere is a filmmaker, health and science journalist, and the author of the *New York Times* bestselling book *Genius Foods: Become Smarter, Happier, and More Productive While Protecting Your Brain for Life*, which has been published in eight languages as of this writing. He is the director of the film *Bread Head*, the first-ever documentary about dementia prevention through diet and lifestyle, and is also the host of the #1 iTunes health podcast *The Genius Life*. Lugavere appears regularly on *The Dr. Oz Show*, *The Rachael Ray Show*, and *The Doctors*. He has contributed to *Medscape*, *Vice*, *Fast Company*, CNN, and the *Daily Beast*; has been featured on *NBC Nightly News*, the *Today Show*, and in the *Wall Street Journal*. He is an internationally sought-after speaker and has given talks at South by Southwest; the New York Academy of Sciences; the Biohacker Summit in Stockholm, Sweden; and many others.